ROUTLEDGE LIBRARY EDITIONS: COLD WAR SECURITY STUDIES

Volume 56

THE SOVIET UNION AND NORTHERN WATERS

THE SOVIET UNION AND NORTHERN WATERS

Edited by
CLIVE ARCHER

LONDON AND NEW YORK

First published in 1988 by Routledge

This edition first published in 2021
by Routledge
2 Park Square, Milton Park, Abingdon, Oxon OX14 4RN

and by Routledge
605 Third Avenue, New York, NY 10017

Routledge is an imprint of the Taylor & Francis Group, an informa business

© 1988 Clive Archer

All rights reserved. No part of this book may be reprinted or reproduced or utilised in any form or by any electronic, mechanical, or other means, now known or hereafter invented, including photocopying and recording, or in any information storage or retrieval system, without permission in writing from the publishers.

Trademark notice: Product or corporate names may be trademarks or registered trademarks, and are used only for identification and explanation without intent to infringe.

British Library Cataloguing in Publication Data
A catalogue record for this book is available from the British Library

ISBN: 978-0-367-56630-2 (Set)
ISBN: 978-0-367-56083-6 (Volume 56) (hbk)

Publisher's Note
The publisher has gone to great lengths to ensure the quality of this reprint but points out that some imperfections in the original copies may be apparent.

Disclaimer
The publisher has made every effort to trace copyright holders and would welcome correspondence from those they have been unable to trace.

ISBN: 978-0-367-56085-0

THE SOVIET UNION AND NORTHERN WATERS

THE SOVIET UNION AND NORTHERN WATERS

Edited by
CLIVE ARCHER

ROUTLEDGE
London & New York
for
THE ROYAL INSTITUTE OF
INTERNATIONAL AFFAIRS
London

First published in 1988 by
Routledge
a division of Routledge, Chapman and Hall,
11 New Fetter Lane, London EC4P 4EE

Published in the USA by
Routledge
a division of Routledge, Chapman and Hall, Inc.
29 West 35th Street, New York NY 10001

© 1988 Clive Archer

Typeset by LaserScript Limited, Mitcham, Surrey

All rights reserved. No part of this book may be reprinted or reproduced or utilised in any form or by any electronic, mechanical, or other means, now known or hereafter invented, including photocopying and recording, or in any information storage or retrieval system, without permission in writing from the publishers.

British Library Cataloguing in Publication Data

The Soviet Union and northern waters.
    1. North Atlantic region. International security. Implications of military policies of Soviet Union 2. Soviet Union. Military policies. Implications for international security of North Atlantic region
    3. Soviet Union. Military policies, 1945-1972
    I. Archer, Clive
    327' .1'16'091821
ISBN 0-415-00489-6

Contents

List of Tables	vii
List of Figures	ix
Abbreviations and Acronyms	xi
Contributors	xv
Preface	xvii
1. Introduction *Clive Archer*	1
2. The Soviet Union in Northern Waters — Implications for Resources and Security *Finn Sollie*	13
3. The Soviet Union and Jurisdictional Disputes in Northern Waters *Robin Churchill*	44
4. The Maritime Policies and Practices of the Soviet Union after UNCLOS III, with Special Reference to Northern Waters *Uwe Jenisch*	62
5. Soviet Fisheries and Offshore Exploration in the Barents Sea *David Scrivener*	76
6. Soviet Military Strategy and Northern Waters *Tomas Ries*	90
7. A Naval Force Comparison in Northern and Atlantic Waters *Robert van Tol*	134
8. The Nordic Response to the Soviet Presence *Clive Archer*	164
9. Responding to the Soviet Presence in Northern Waters: An American Naval View *Douglas Norton*	179
10. The Maritime Strategy and Geopolitics in the High North *Steven Miller*	205
11. Maritime Strategy in Northern Waters: Implications for the Navies of Europe *Geoffrey Till*	239
Index	252

Tables

5.1	Main Soviet nominal catches (by species) in the North-east Atlantic Fishing Area, 1982	*77*
5.2	Percentage take of total nominal catch of the most important species in the Svalbard-Bear Island Fishing Area, 1981 and 1982	*81*
6.1	The translation from major strategic objective to strategic operational task	*112*
6.2	The translation from strategic operational task to regional strategic support requirements	*113*
6.3	The translation from regional strategic support requirements to specific northern TVD strategic support missions	*114*
7.1	Soviet Northern Fleet/NATO's Atlantic naval forces: the aggregate statistics	*135*
7.2	Soviet Northern Fleet/NATO's Atlantic naval forces by basing areas	*137*
7.3	Peacetime deployment of the Soviet Northern Fleet and NATO's northern naval forces	*138*
7.4	Presumed initial wartime deployment of the Soviet Northern Fleet and NATO's northern naval forces	*139*
7.5	Ship and submarine class age	*141*
7.6	NATO and Soviet naval construction, 1976-85	*142*
7.7	The number of vessels built to new designs	*142*
7.8	New vessels joining the Northern Fleet as a percentage of the total	*144*
7A.1	The composition of the Red Banner Northern Fleet, 1975-85	*148*
7A.2	The composition of forces associated with the Kola Peninsula	*150*
7A.3	NATO naval forces available for operations in the Norwegian Sea, North Sea (excluding the English Channel), Barents Sea and North Atlantic	*152*
7A.4	Soviet Northern Fleet/NATO's Atlantic naval forces by basing areas	*154*

TABLES

7A.5 Soviet Northern Fleet/NATO's Atlantic naval forces by
presumed initial wartime and peacetime operating areas *155*

7A.6 The age of submarine and ship classes design: the age of
warship and submarine class design, as of 1/1/86, by
hull-life from the date of the commissioning of the first
unit of a class *158*

Figures

3.1	Disputed areas in the Barents Sea	*46*
6.1	The evolution of Soviet strategic delivery vehicles, 1960–84	*93*
6.2	Soviet SSBN basing and launch areas, 1959–68	*95*
6.3	Soviet SSBN basing and launch areas, 1968–73	*97*
6.4	Soviet SSBN basing and major launch areas, 1973–85	*98*
6.5	Major flight paths of US airborne strategic systems to Soviet counterforce targets, 1960	*101*
6.6	Strategic airborne transit routes to Soviet counterforce targets, 1985	*102*
6.7	Strategic airborne transit routes to Soviet countervalue targets, 1960	*103*
6.8	Strategic airborne transit routes to Soviet countervalue targets, 1985	*104*
6.9	The evolution of Soviet northern naval operations for defence against the seaborne nuclear strategic threat	*106*
6.10	Soviet military perception of the Northern Waters area	*111*
6.11	Soviet northern air defence operations	*118*
6.12	Soviet northern naval defence operations	*119*
6.13	Air control using only Soviet bases	*122*
6.14	Air control using North Norway	*123*
6.15	Air control using North Norwegian Sea	*124*
6.16	Air control after attack on Norwegian bases	*125*
6.17	Air control using Scandinavian bases	*126*
6.18	Probable main thrust of Soviet ground operations in wartime in the northwestern TVD	*127*

Abbreviations and Acronyms

AASU	Air Army of the Soviet Union
ABM	Anti-ballistic missile
ACE	Allied Command Europe
AD(S)	Air defence (sector)
AEW	Airborne early warning
AFNORTH	Allied Forces Northern Europe
AJIL	*American Journal of International Law*
ALCM	Air-launched cruise missile
AMF	Allied Command Europe Mobile Force
ASW	Anti-submarine warfare
AWACS	Airborne warning and control system
BAM	Baikal–Amur–Magistrale (Railway)
BMD	Ballistic missile defence
BMEW(S)	Ballistic missile early warning (system)
CAST	Canadian Air–Sea Transportable (brigade)
CBO	Congressional Budget Office (USA)
CENTLANT	Central Atlantic Area (of NATO's Allied Command Atlantic)
CGN	Guided missile cruiser (nuclear)
CINCHAN	Commander-in-Chief Channel
CNO	Chief of Naval Operations
COB	Collocated operating base
COCOM	Co-ordinating Committee for Multilateral Export Controls
CONMAROPS	Concept of Maritime Operations
DEW	Distant early warning
EEZ	Exclusive Economic Zone
EW	Early Warning
FAC	Fast attack craft (gun)
FPZ	Fisheries Protection Zone
FRG	Federal Republic of Germany (West Germany)
GDR	German Democratic Republic (East Germany)
GI(F)UK	Greenland–Iceland(Faroes)United Kingdom (Gap)
GTVD	Main theatre of military operations
GYIL	*German Yearbook of International Law*
IA	Interceptor aviation

ABBREVIATIONS AND ACRONYMS

IAPVO	Interceptor aviation of the air defence forces
ICBM	Intercontinental ballistic missile
ICES	International Council for the Exploration of the Sea
ICJ	International Court of Justice
IDR	*International Defense Review*
IISS	International Institute for Strategic Studies
INF	Intermediate-range nuclear forces
JDW	*Jane's Defence Weekly*
LOS	Law of the Sea
LRB	Long-range bomber
LRTNF	Long-range theatre nuclear forces
MA	Maritime aviation
MAB	Marine Amphibious Brigade (USA)
MARCONFORLANT	Maritime Contingency Force Atlantic
MCM	Mine countermeasures
MD	Military district
MIRVS	Multiple warheads *or* Multiple independently targeted re-entry vehicles
MIZEX	Marginal Ice Zone Experiment
MR	Medium range
MRB	Medium-range bomber
NAEW	NATO Airborne Early Warning Force
NAEW & C	NATO's Early Warning and Control System
NATO	North Atlantic Treaty Organization
NAVOCFORMED	Standing Naval Force Mediterranean
NNFZ	Nordic Nuclear-free Zone
NORCCIS	Joint Command, Control and Information Systems
NORLANT	Northern Sub-area, Eastern Atlantic Area (of NATO's Allied Command Atlantic)
ORBAT	Order of Battle
OTVD	Ocean theatre of military operations
PVO	Air defence
RAF	Royal Air Force (UK)
RPV	Radar and electronic warfare troops
RRP	Rapid Reinforcement Plan
RUSI	Royal United Services Institute for Defence Studies
SAC	Strategic Air Command
SACEUR	Supreme Allied Commander Europe
SACLANT	Supreme Allied Commander Atlantic

ABBREVIATIONS AND ACRONYMS

SAM	Surface-to-air missile
SDI	Strategic defense initiative
SLBM	Submarine- (or sea-) launched ballistic missile
SLCM	Submarine-launched cruise missile
SLEP	Service life extension programme
SLOC	Sea line of communication
SONOR	South Norway Command
SOSUS	Sound ocean surveillance system
SSB	Submarine with ballistic missiles
SSBN	Nuclear-powered submarine with ballistic missiles (or strategic submarine)
SSGN	Nuclear-powered submarine with cruise missiles
SSN	Nuclear-powered submarine with torpedoes
STANAVFORCHAN	Standing Naval Force Channel
STANAVFORLANT	Standing Naval Force Atlantic
TAC	Total allowable catches
TVD	Theatre of military operations
UK/NL AF	United Kingdom-Netherlands Amphibious Force
UNTS	United Nations Treaty Series
USNIP	*United States Naval Institute Proceedings*
VMF	Navy (Soviet Union)
VVS: DA	Long-range aviation of the airforce
WESTLANT	Western Atlantic Area (of NATO's Allied Command Atlantic)
ZRV	Air defence (missile) troops

Contributors

Clive Archer	Centre for Defence Studies, University of Aberdeen
Robin Churchill	University of Wales Institute of Science and Technology, Cardiff
Uwe Jenisch	Ministry for Economics, Land of Schleswig-Holstein, Kiel
Steven Miller	Massachusetts Institute of Technology, Boston
Douglas Norton	US Navy, US Mission to NATO, Brussels
Tomas Ries	Norwegian Institute for Defence Studies, Oslo
David Scrivener	Department of Politics and International Relations, University of Aberdeen
Finn Sollie	The Northern Perspectives Group, Oslo
Geoffrey Till	Department of History, Royal Naval College, Greenwich
Robert van Tol	Royal United Services Institute for Defence Studies, London

Preface

In 1986 Croom Helm published, for the Royal Institute of International Affairs, *Northern Waters: security and resource issues,* which included a number of contributions from the Northern Waters Study Group of the Scottish Branch of the Royal Institute. This Study Group brought together academics, business men, civil servants and serving officers interested in Northern Waters and helped arrange a number of seminars and international conferences. Its members also had contacts with those in Scandinavia and North America who had a professional involvement in Northern Waters.

Since the establishment of the Study Group in 1979, interest in Northern Waters has flourished in Britain, the United States, Canada, West Germany and the Nordic countries. In Autumn 1985 the Centre for Defence Studies, University of Aberdeen, held an International Colloquium on what have probably been the main inspirations for the attention devoted to Northern Waters — increased Soviet activity therein and the response of the Western powers. This book reflects some of the issues dealt with at that colloquium and, like the 1986 book, covers jurisdictional and resource questions as well as those concerned with international security.

I am grateful for the support given for the International Colloquium and the subsequent publication of this book from the Binks Trust, the Ford Foundation and the Norwegian Atlantic Committee. Help and encouragement were received in both endeavours from David Greenwood and David Scrivener. The work of Margaret McRobb in typing successive drafts of the chapters is particularly appreciated.

Clive Archer
Aberdeen

Frontispiece: The Soviet Union and Northern Waters

1

Introduction

Clive Archer

Northern Waters: security and resource issues[1], dealt with a particular area of the North Atlantic, touching on legal questions, resources, transportation, strategic factors — including the effect of new military technology and the problems of controlling conflict — and identifying the interests of the states in and bordering the region. Northern Waters were defined as the maritime areas within the latitudes of 80°N and 60°N and from longitude 90°W to 40°E. This sector includes the islands of Arctic Canada, Greenland, Iceland, the Faroes, Shetlands, Jan Mayen and Svalbard and extends to the Kola Peninsula in the north-west of the Soviet Union.

As a medium of communication between areas whose strategic value has increased in the last 25 years, Northern Waters have assumed greater significance for the Soviet Union as well as for the United States. The area has three important approaches: the Fram Straits between Greenland and Svalbard, joining the Polar Sea to the Atlantic Ocean; the Baffin Bay/Davis Strait waters which link the northern coasts of Alaska and Canada to the Atlantic Ocean; and the Barents and Norwegian Seas which provide an egress from the northern regions of Norway and the Soviet Union and from the Arctic Ocean to the North Atlantic. Northern Waters have been of particular interest to the Soviet Union because of its need for access to the North Atlantic — and the Arctic — from its massive military base areas on the Kola Peninsula.

Furthermore, these waters have an intrinsic value by virtue of economic resources such as fish, natural gas and oil. Until the 1970s these resources existed for the most part in the commons of the high seas. From the mid-1970s both fisheries and mineral exploitation have become increasingly encompassed within the national jurisdiction of the littoral states which have created exclusive fisheries and economic zones.[2] For the Soviets this has meant a possible reduction in direct access to these resources and a need either to contest control of certain areas (for example, the sea border between Norway and the Soviet Union) or to reach agreements with other littoral states over use of resources.

This book is about the presence of the Soviet Union in Northern Waters. After a general overview (Chapter 2), it covers the legal aspects (Chapters 3 and 4), resource questions (Chapter 5) and the strategic implications of that presence (Chapters 6 and 7). Also examined are the responses of the major Western states with interests in the area (Chapters 8, 9, 10 and 11).

The heavy emphasis on the reaction of the West to the Soviet presence needs some explanation. It can be seen as a purely occident-centred response: Western interest in Soviet activities tends to stress the effect on its own sphere rather than the outcomes for the population of the USSR. After all, a large part of these Northern Waters represents a link between two of the most important areas of the West: North America and Western Europe. It is also dealing with what is known in the West: Soviet intentions are still difficult to decipher and their capabilities — economic as well as military — are not always fully understood, whereas those on the Western side tend to be more open to scrutiny. Finally, there is an intellectual side to this bias which should be considered. As can be seen from these chapters, a number of Western activities in Northern Waters are affected by, and others explained by, the Soviet presence and its growth in recent decades.

Has it been the case that certain West European and North American activities have resulted from the Soviet expansion into Northern Waters? Clearly no such causation can be claimed for many events such as the extension of economic and fisheries zones and for what could be regarded as the precautionary deployment of Western maritime assets to the area. The latter would no doubt have taken place had Soviet air and sea forces remained bottled up in their home ports as they had done for much of the 1940s and 1950s. What the expanded Soviet presence has meant, *prima facie*, is an increased value for the resources of the area to which the Soviets might seek access through a range of economic arrangements and legal settlements. A withdrawn Soviet Union demonstrates either lack of willingness or of capability to involve itself in the Northern Waters' market-place wherein jurisdictional rights and resource accessibility are traded between governments and enterprises. A Soviet presence in a sense provides an extra and potentially important 'customer', thereby increasing the value of Northern Waters to the other littoral states. This presence has therefore increased the opportunities for Western countries in Northern Waters by broadening the range of relationships they can have there, and has also affected the willingness of certain governments — most noticeably that of the United States — to become involved in the area in a particular fashion, mainly that of a naval presence.[3]

This propensity for increased US involvement has not only been occasioned by the expanded Soviet Northern Fleet. It has also been made possible by the changed political situation in the United States of the 1980s that has brought with it an expanded defence budget and a naval leadership with the determination and political ability to formulate the Forward Maritime Strategy.[4] A greater presence — both civil and military — in Northern Waters has been made easier by technological change over the last ten to 15 years. Advances in ice-breakers, drilling in northern regions, under-ice activity and in satellite communications have been particularly important for the region and have allowed activity there on a scale not contemplated beforehand.[5]

Thus, it is too simplistic to say that an increased Soviet maritime presence in Northern Waters has *caused* an equivalent Western build-up. Soviet activity and technological developments have evoked a response from a more willing US Administration and perhaps more aware West European governments. What can be said is that the decision-makers in the relevant Western countries — the United States, Canada, Denmark, the Federal Republic of Germany, Great Britain, Iceland, the Netherlands, and Norway — have perceived an increased threat to their interests in Northern Waters and have responded either by increasing their national defence capacity there or by trying to decrease the national defence risk or by a combination of the two actions.[6]

The first response is reflected in the more prominent position of the US Navy in Northern Waters over the last few years and in the development of a more robust defence posture there by West Germany, Iceland, Norway and the Netherlands. The second strategy is that of *détente*, pursuing a lessening of tension in relations with the Soviet Union, which was a dominant strand in West German foreign policy in the 1970s and can still be seen as an important element in the dealings by that country — and others such as the United Kingdom — with East European countries. The final option — that of a deliberate mixture of increased defence capacity with a conscious effort to decrease the defence risk — has been followed by the Nordic members of NATO since 1949. They have sought to 'buy in' extra security capacity by their membership of NATO and by the extension of the

Atlantic Alliance's collective defence and nuclear weapon-based deterrent. They have balanced this by an effort to reassure the Soviet Union that military activities in their countries will be restricted in such a way as not to threaten the main strategic resources of the Soviet Union.[7] The reassurance element covers more than just military aspects: it also involves a willingness to treat with the Soviet Union as a valid participant in the international decision-making process in legal and economic matters in that area.

A 'mixed' policy by the Western states towards the Soviet presence in Northern Waters is not an easy one to implement. Firstly, one aspect of the policy might come under attack domestically, leaving an unbalanced approach. To a certain extent this happened to President Carter's arms control policy at the end of his term of office: the opposition to the Strategic Arms Limitation Treaty II turned the US approach to the Soviet Union into one that stressed armed strength, even before the election of President Reagan. An example of the opposite happening can be found in Denmark in the mid-1980s. The precarious parliamentary situation meant that the right–centre government of Mr Schlüter was unable to follow the sort of defence policy it preferred and to get the level and type of military expenditure it wanted to match the Danish policy of reassurance of the Soviet Union.[8] Secondly, because the main Western states involved in Northern Waters are North Atlantic Treaty Organization (NATO) members, there may be problems in running a co-ordinated Alliance policy. If some states decide on a 'mixed' approach of deterrence and reassurance towards the Soviet Union whilst others show a preference for a robust defence stance, then the signals to the Soviet Union will be confusing and the effectiveness of any 'NATO' policy will be reduced. There could be a danger that elements of the American Forward Maritime Strategy could impinge on Norwegian home waters in such a way that the Soviets might consider it as counteracting Norwegian unilateral military restraints in pursuit of their policy of reassurance.[9] Thirdly, a subtle policy of deterrence and reassurance could be futile if misunderstood — or not understood at all — by the Soviet Union. Finally, there is the problem of negative feedback. A Western emphasis on achieving a legal settlement to a dispute, for example, may encourage a Soviet perception of its adversaries' weakness — and a consequent tough response — which could in turn lead to retrenchment by the Western powers. As Johan Jørgen Holst and others have written, deterrence can slip into provocation and reassurance into appeasement. In short, any 'mixed' Western policy in this area should, ideally, have the features of coherence, consistency, and clarity. It should also call forth the predicted positive response from the Soviet Union: caution in the face of deterrence, contentment in response to reassurance.

The chapters in this book demonstrate the difficulties of following a 'mixed' approach towards the Soviet presence in Northern Waters. They portray the reserved approach by most Western governments to most relations with the Soviet Union in this area. They also show different interpretations of the relevance of the Soviets' northern policies and the appropriateness of Western responses. In security terms, there is not always agreement as to whether, in dealing with the Soviet Union, the priority should be with increasing the Western defence capacity or with decreasing the defence risk.

The jurisdictional disputes between the Soviet Union and Norway over parts of the Barents Sea and the waters around Svalbard are dealt with in two chapters in this book. There is little disagreement with Robin Churchill's contention that as long as the disputes continue — though at a low intensity — there will be less than optimum management of the resources of the area. However, there is less agreement about possible solutions to the outstanding jurisdictional differences between the Soviet Union and Norway. Whilst Uwe Jenisch suggests that the Soviet claims in the Arctic 'can no longer be taken seriously', Robin Churchill gives them some credence in the case of the division of the Barents Sea, based on the greater length of the Soviet coastline compared with that on the Norwegian side. Even this approach does not accept the Soviet idea of the sector line dividing the Norwegian and Soviet parts of the Barents Sea.

The question about the Svalbard (Spitsbergen) Treaty is whether it extends to a 200-mile zone around the archipelago. If so, as the Soviets claim, then the economic regime therein would be subject to different regulations than in the rest of Norwegian waters, as the Treaty allows relatively free access to the resources of Svalbard for the signatory powers. The Norwegian government has claimed that the area outside the territorial waters of Svalbard is not covered by the Treaty and that normal Norwegian law applies to that zone. Robin Churchill is not convinced by the Norwegian claims, though neither does he find the Soviet case conclusive. He argues for a diplomatic solution. Uwe Jenisch starts from a different premiss: he does not consider that the Svalbard Treaty covers the continental shelf and Exclusive Economic Zone of the archipelago and, seeing the danger of disorder if a free-for-all is allowed, he calls for a political solution. Though the two writers see the Soviet–Norwegian maritime disputes in different lights, both of them accept that a mutual settlement is preferable. While this would not allow the Soviet Union the pre-eminent position it seems to expect in this area, a solution might offer the Soviets enough of an interest — without giving them *droit de regard* — to encourage the management of the resources there.

The current situation with regard to the utilisation of resources in Northern Waters, as outlined in the chapters by Finn Sollie and David Scrivener, is one of unfulfilled potential. The jurisdictional disputes around Svalbard and in the Barents Sea, referred to above, provide a partial explanation for this state of affairs. Any oil and gas reserves in Northern Waters have been in inhospitable waters and have had to await the appropriate technology — and favourable market conditions — for exploitation. Though an increased Soviet presence in Northern Waters may enhance the political and economic value of these resources, this will not be the case should other countries be excluded from the area by Soviet activity. In Chapter 5, David Scrivener alludes to the bilateral element creeping into Soviet dealings with Norway over the resources on and around Svalbard and in the Barents Sea. This has been resisted by Norwegian governments, which have disputed not only the legal basis of Soviet claims but also the political reasoning behind them. Norway has not wanted to be isolated from its Western allies in economic and security matters. A condominium would not only lower the economic value of the areas concerned but would provide a dangerous precedent for Soviet involvement in the internal affairs of a Western country.

This does not mean that co-operative arrangements for Northern Waters' resources cannot be negotiated between the Soviet Union and Western countries. Attempts have recently been made to revive the historical experience of the Russian–Scandinavian 'Pomor' trade in the Arctic Circle areas. A revival clearly depends on economic factors such as the price of oil but it also needs a favourable political climate. Neither the Soviet Union nor Norway is likely to initiate a close trading relationship in these northern areas in a climate of tension or whilst the adjacent seas are so militarily sensitive and their maritime frontiers remain undelineated. Whatever commercial agreement is reached over the exploration and exploitation of resources, its effect on security policy would be uncertain. It could be seen as part of a 'mixed' approach, mentioned above, towards the Soviet Union, by which the Soviets are given a vested interest in the further peaceful development of the northern region. However, it is unlikely that too much political significance will be attached to East–West commercial agreements in the area, not least because of the wariness born of the experience with *détente* in the 1970s when such linkage proved elusive.

In considering the security side of the equation of Soviet presence in Northern Waters, a vital question is its nature and size. In Chapter 6, Tomas Ries divides Soviet strategy in Northern Waters into that involving the strategic forces, which has an extra-regional orientation, and that related to the consequent theatre-level support requirements which have a regional direction and thereby 'decisively affect the local military equilibrium'.[10] The first aspect, according to Ries, determines the level of superpower strategic interest in the area, while the latter shapes the objectives and scope of military operations at theatre level in the region. This chapter traces the five major reasons for Soviet strategic interest in its north-west and adjacent areas — sea-launched strategic nuclear attack, defence against a seaborne nuclear threat, defence against an airborne nuclear threat, airborne strategic nuclear attack, and cutting NATO's Atlantic sea lines of communication (SLOCs) — and examines the regional effect of forces allocated to safeguard these strategic interests. Ries concludes that while Soviet strategic activity in Northern Waters does not directly affect the regional equilibrium of forces, it does help to determine the overall stability of the area by impinging on the Soviet theatre force-levels and by increasing tension. Northern Waters themselves may not be a primary strategic objective, but their importance to the Soviet Union has grown over the past 30 years and is likely to increase. Ries notes the efforts by some in the Nordic area — in the form of Nordic nuclear-free zones or plans for strategic submarine (SSBN) sanctuaries — to try to decrease the risk to their countries' national defence of the superpower involvement in Northern Waters.

In examining the support and safeguard systems for the strategic forces that the Soviets have in their north-western areas, Ries contends that these increasingly affect the regional military equilibrium but are primarily defensive. The main response of the Nordic states to this perceived threat has been to increase their national defence capacity but not in such a way that could trigger a change from a defensive to an offensive mode in Soviet forces.

Looking at the tactical forces in those areas of the USSR bordering the Northern Waters, Ries suggests that a certain amount of mutual restraint seems to have taken place between the Nordic states and the Soviet Union. This implies that the 'mixed' approach might have worked at this level. If so, its continued success needs constant vigilance.

In Chapter 7 Robert van Tol compares the naval strength of the Soviet Union in Northern Waters with that of the Western powers. This calls into question the extent of the 'Soviet threat' to which the West is responding. Van Tol typifies the Soviet Northern Fleet as having limitations and problems, most noticeably those associated with small, ageing and unsophisticated vessels. However, the Fleet does have certain strengths — when NATO forces are dispersed by region, the Soviet Union could control the Norwegian and Greenland Seas if NATO forces failed to move northwards. This strength is far from overwhelming and even the forces of small Norway could cause the Soviet military considerable trouble. Furthermore, NATO forces have a geographical advantage in restricting the activities of the Northern Fleet.

Van Tol's chapter shows that, in time of conflict, Soviet military leaders would face difficult decisions about their forces' deployment in Northern Waters. A major problem would involve the placing of submarines south of the GIUK Gap: the more committed to this duty, the less the Soviets would have to defend their strategic submarines and to keep control of the Norwegian Sea. Van Tol identifies sections of the Norwegian Sea as being crucial to both NATO and the Soviet Union in any conflict and it is his contention that, given the necessary resources, NATO could prevail in the southern and central part of that sea, but land-based cover could well decide how far north control could be exercised. Van Tol concludes that the 'highly publicised Soviet submarine threat is in fact equalled, quantitatively, and bettered, qualitatively, by NATO'.[11] This suggests that the 'Soviet threat' in Northern Waters has been somewhat overdrawn and that the Western countries bordering the area may have a greater freedom to adjust the elements of a 'mixed' policy towards the Soviet Union than is normally considered.

The final four chapters examine the Western response to the Soviet maritime presence in and around Northern Waters. This author's chapter deals with both the political and military reaction of the Nordic states, especially the NATO members. As noted earlier, these countries have from an early stage attempted a 'mixed' policy of reassurance and deterrence and have experienced some of the associated difficulties. With the effect at a local level of strategic factors (as noted by Tomas Ries), the Nordic countries have found themselves increasingly responding to the international security environment with only a limited ability to mould their milieu. There is even a danger that steps taken actively to change their situation could have a negative effect. Thus, for example, too great an experiment with elements of reassurance (such as nuclear-free zone proposals or further military restriction) could lead to Allied reluctance to reinforce Norway or Denmark in times of conflict. What happens to these countries is of interest not just because they represent NATO's 'Northern Flank': their policies of restraint — no nuclear weapons, no foreign bases, curbs on Allied exercises — reflect what could be the policies of future mainland West European governments. Whether these approaches are, or have been, successful in improving the security situation of the Nordic countries may provide indicators as to their success elsewhere. As yet, the verdict is the Scottish one of 'not proven'.

The crucial country in any consideration of a Western response to the Soviet presence is the United States. The chapters by Steven Miller and Douglas Norton look at America's Maritime Strategy, especially as it affects Northern Waters. These two different perspectives agree on certain factors. Both see the Maritime Strategy as one aimed at deterring the Soviet Union and only concerned with war-fighting once deterrence has failed. They consider that the Strategy places a premium on an early recognition of a crisis and neither considers that the United States is contemplating a 'smash and grab' action against the Kola Peninsula or has a fixed timetable for its military operations in case of conflict in Northern Waters. In examining the Maritime Strategy's genesis, the emphasis of each author is different. Norton stresses the historical background while pointing out new additions, while Miller admits elements of continuity but considers that the openness of discussion is novel, as are certain aspects of the plans. Norton also sees the Strategy as dealing operationally with a number of elements in an integrated fashion but Miller points to the differing requirements of the Strategy.

The two authors have different evaluations of the US Maritime Strategy as a fitting response to Soviet military strength in Northern Waters. Miller questions what he and other writers consider the wreckless elements of the Strategy, in particular operating aircraft carriers so close to Soviet bastions, while Norton emphasises cautious and contingent aspects of the Maritime Strategy. He points to the slow-moving nature of naval operations and in his chapter the Norwegian Sea rather than the Barents Sea, which is bound to be more hazardous, receives a good deal of attention. The discussion in the two chapters about the effect of any US threat to Soviet SSBNs links up with findings of Tomas Ries and Robert van Tol that in any naval conflict in Northern Waters, the Soviet Union is likely to concentrate its efforts on protecting its strategic assets. Some have seen this as part of the rationale for the US Maritime Strategy in the area: by placing Soviet SSBNs in danger, it forces the Soviets to stay in the far north and does not provide them with the easy option of dedicating submarines to the destruction of NATO's SLOCs. Critics point to the risks involved in such a policy: it could force the Soviets to 'go nuclear' at an earlier stage in the conflict and it might evince the response of an attack on the SLOCs precisely to draw off American pressure on the Soviet northern bastions.

In the end, much of the debate about the Maritime Strategy may be premature, especially if in future the resources are not available to allow its full implementation. Furthermore, much of the speculation is dependent on the Soviet response to the strategy, and little is known about that as yet. However, it is essential that such an important aspect of the Western response to the Soviet Union should be discussed at this stage, not least because it has implications for America's European allies. As Geoffrey Till points out in Chapter 11, 'strategic shadow-boxing between the two superpowers concerns the seas off Europe and naturally European governments...have an interest in all these activities'.[12] He contends that the Maritime Strategy is a reflection of American interests and an American view of any war with the Soviets in the North Atlantic area, and he brings out the implications for West European navies. Till reminds us that Northern Waters are not the only likely seas for a superpower confrontation and that the Maritime Strategy has a global aspect which in itself presents certain problems for Europeans.

In the end, Till notes that most of the West European allies welcome the forward deployment of US forces in Northern Waters as it helps them to 'multilateralise' their relations with the Soviet Union by bringing in US assistance. He points to the European wish to match the resulting increased defence capacity with efforts to reassure the Soviets that this does not imply hostile intent. This leads back to the difficulty of following a 'mixed' policy towards the Soviet Union in the area, a motif clear in the chapters of this book. It should perhaps be recognised that the Soviet Union, under the leadership of Mikhail Gorbachev, is now itself pursuing something of a mixed policy towards Western Europe. Military might is being matched by reassurance and by efforts to decrease tension.

In the same spirit, Mr Gorbachev's speech in Murmansk on 1 October 1987 contained proposals for the northern area, emphasising the 'lowering of the level of military confrontation in the region' and 'developing co-operation'.[13] Gorbachev suggested that the Soviet Union would act as a guarantor for a Nordic Nuclear-free Zone and might be prepared to take measures in its own territory to make such a Zone less problematic. He also took up the proposal of President Koivisto of Finland about restricting naval activity in Northern Waters and the Baltic, and proposed confidence-building measures in these areas.

On peaceful co-operation, Gorbachev expressed Soviet interest in joint exploitation with the West of oil and gas on the shelf of the northern seas: he proposed a conference of sub-Arctic states to co-ordinate Arctic research and he suggested a single comprehensive plan to protect the nature of the north.[14] The West's immediate response was either cautious or negative, but indicated that a more detailed consideration might eventually be forthcoming.[15] In particular, the Nordic states will wait to see whether the Soviet Union's apparently new approach to the northern areas filters down to negotiations on joint enterprises, fisheries agreements and the Barents Sea. If the spirit of Murmansk is transferred to Soviet operational behaviour in Northern Waters, then the West may find itself with the same mixed emotions that have typified its feelings about other Soviet diplomatic moves in Europe.

NOTES

1. Clive Archer and David Scrivener (eds), *Northern Waters: security and resource issues* (Croom Helm, for the Royal Institute of International Affairs, London, 1986).
2. Patricia Birnie, 'The Law of the Sea and Northern Waters' in Archer and Scrivener (eds), *Northern Waters*.
3. The concepts of opportunity and willingness in foreign policy analysis have been used by Harvey Starr in '"Opportunity" and "willingness" as ordering concepts in the study of war', *International Interactions*, vol. 4, no. 4 (1978), pp. 363-87, and also in Bruce Russett and Harvey Starr, *World politics: the menu for choice* (W.H. Freeman, San Francisco, 1981).
4. See below, Chapter 9, Douglas Norton, 'Responding to the Soviet presence in Northern Waters: an American naval view'; and Chapter 10, Steven Miller, 'The Maritime Strategy and geopolitics in the High North'.
5. For details of the effects of technological change, see the following papers in Archer and Scrivener (eds), *Northern Waters*: Tony Scanlan, 'Resource endowment and exploitation', pp. 42-54; Terence Armstrong, 'Transportation of resources from and through the Northern Waters', pp. 55-68; and David Hobbs, 'New military technologies and Northern Waters', pp. 85-96. See also Tom Stefanick, *Strategic and antisubmarine warfare and naval strategy* (Lexington Books and Institute for Defense and Disarmament Studies, Lexington, 1987).
6. See Benjamin A. Most and Harvey Starr, 'International relations theory, foreign policy substitutability, and "nice" laws', *World Politics*, vol. XXXVI, no. 3 (1984) p. 393.
7. See Johan Holst, Kenneth Hunt and Anders Sjaastad (eds), *Deterrence and defense in the North* (Norwegian University Press, Oslo, 1985), and Clive Archer, 'Deterrence and reassurance in northern Europe', *Centrepiece 6*,

(Centre for Defence Studies, Aberdeen, Winter 1984).

8. Martin Heisler, 'Denmark's quest for security' in Gregory Flynn (ed.), *NATO's northern allies* (Croom Helm, London, 1985), and Clive Archer, 'The political and economic context', paper delivered at conference on Britain and the Security of NATO's Northern Flank, King's College, London, May 1986.

9. For this sort of scenario, see Barry Posen, 'Inadvertent nuclear war? Escalation and NATO's Northern Flank' in Steven E. Miller (ed.), *Strategy and nuclear deterrence* (Princeton University Press, Princeton, 1984), pp. 96–106.

10. See Chapter 6.

11. See Chapter 7.

12. See Chapter 11.

13. Novosti Press Agency, 'Gorbachev-Speech in Murmansk', *Press Release PR62887*, 2 October 1987, pp. 2 and 10.

14. As note 13, pp. 8-10.

15. See, for example, the *Guardian*, 9 October 1987, p. 6; *Statement by the Minister for Foreign Affairs Mr Sten Anderssen*, Ministry for Foreign Affairs, Press Release, Stockholm, 2 October 1987; *Aftenposten*, 2 October 1987, p. 8; *Statsministerens Kommentar til Mikhail Gorbatsjovs tale i Murmansk 1 Oktober*, Statsministerens Kontor, Pressmelding, Oslo, 2 October 1987.

2

The Soviet Union in Northern Waters — Implications for Resources and Security

Finn Sollie

RUSSIA GOES TO SEA

In the last few decades, the Soviet Union has emerged as a major maritime power. It has a large merchant marine which keeps growing in apparent disregard of business cycles; it has the world's largest fishing fleet in numbers and competes with Japan for the largest catch; it has a larger fleet of ocean research vessels than any other country; and, most importantly of all, it has a navy of conventional ships and of attack and strategic submarines which is second to none in Europe and a formidable rival to the US Navy in the Atlantic, as well as in the Pacific Ocean. In short, the government of the Union of Soviet Socialist Republics has carried out a maritime development programme far beyond that which the great Tsar Peter could have hoped for even in his wildest dreams, when he decided to build a new capital (St Petersburg, 1712) on the sea-shore in an effort to open land-bound Russia to sea and to the opportunities of maritime operations.

Russia's new role as a global maritime power coincides with the Soviet Union's rapid development as a global superpower. With Russia ascending to become 'the other superpower', alone with the United States in the superior league, the international power structure has been completely transformed in less than 30 years. The process was helped along by the equally rapid decline of the other great powers and the abolition of the colonial empires of Britain and France, and new weapons, notably nuclear arms to be delivered over long distances, have created a new strategic situation with the two superpowers facing each other directly across the frozen Arctic. In the balance between them, navies, notably strategic submarines with long-range nuclear weapons and countermeasures to neutralise them, play a major and increasing role.

In this new context of an altered power structure and new weapons, old tensions and conflicts of interest remain as a potential threat to peace, and security in Europe continues to be a major international concern. Experience from two world wars and the continuing split of Europe into opposing blocks make it imperative to maintain a naval capability to protect transatlantic supply lines (and conversely, for the other side, to cut them). Thus, there is a double need for the leading powers (and their allies) to maintain a high capability and readiness for war at sea, both as an integral part of any major war of a conventional/traditional nature where sea lanes are important, and as a 'back-up' capability to deter or to threaten nuclear war between the superpowers.

For obvious geographical reasons, the Soviet Union has had to concentrate its naval bases in northern seaports and to use Northern Waters as its main training area and access to the open seas. For the same reasons, great effort is made to develop techniques for deployment and protection of strategic submarines with long-range weapons in near Northern Waters. At first, this concentration took place in the European north, with bases on Kola, but more recently the Soviet Pacific Fleet has been substantially upgraded. This is an important part of the general increase in the importance and role of the Pacific in strategic as well as in other contexts and adds a new centre of strategic gravity to the international maritime balance, but it does not detract from the importance of the European Northern Waters. Any hope that further naval build-up in the Pacific may give relative relief of risk and pressure in the North Atlantic is muted by awareness that this will require dispersion of countermeasures and, of course, by certainty that open confrontation between the superpowers in the Pacific theatre must spread to the Atlantic as well.

The very rapid development of the Soviet Navy, and of its other maritime capabilities, is a natural and, in Russian eyes, a necessary consequence of the Soviet Union's ascendence to the role of leading military power in Europe, and to the status of equal to the United States on the global level. For others, and particularly those of us who feel and fear the threat of the Soviet Union's military strength, it is equally natural to focus attention on the rapid growth in the size, quality and capability of the Soviet Navy as a new danger that did not exist a few decades ago — and, of course, to look for ways and means to counter and to offset the maritime threat.

Soviet naval build-up and the threat from the Northern Fleet is the immediate cause for the great attention now given to developments in European Northern Waters and for the debate on a strategy to counter the naval threat. Discussion of these issues is important and indeed vital for security in the Atlantic and in Europe, as well as for the global balance. It must be noted, however, that the Russian maritime challenge is much more than a temporary military imbalance that can be remedied by revised strategies and military countermeasures. The Soviet Navy will be a formidable enemy in war, but even now, in peacetime, the Soviet Union is already a worthy adversary, using all the maritime capabilities of its sea power to promote its interests and to increase its influence world-wide.

Russia has indeed gone to sea to engage in all types of maritime operation. This is a major new development for Russia itself and one that will have long-term implications for the Soviet Union's role and involvement in international affairs generally and for international maritime activities in particular. To understand the full importance of Russia's new role as a maritime nation, it is necessary to see the development of the sea power of the Soviet state in a wider perspective, against the backdrop of Russian tradition and in the light of Soviet concepts of sea power and maritime operations.

As already noted above, it has been necessary for the Soviet Union to concentrate a major portion of its naval forces in the north, with bases on the Kola Peninsula and Europe's northernmost waters as a training, passage and deployment area. Thus, these waters have gained special importance for the Soviet Union, as well as for the Western Allies as Russia's Atlantic adversaries. We still have not been able to adjust fully to this new situation on the Northern Flank — now often dubbed 'the Northern Front' — either in thinking or in measures to meet the challenge and to deal with the problems. In this context it is important to remember also that the far north is not, and indeed never has been, a region entirely beyond the compass and reach of other maritime nations.

Beginning with early maritime exploration (sixteenth century), European maritime states have engaged in operations in Northern Waters for hundreds of years, and with the advent of coal-mining on Spitsbergen (c. 1900), US interests also became involved. The Spitsbergen Treaty of 1920 recognised Norway's 'full and absolute sovereignty' over the islands, but subject to 'stipulations' (art. 1) to secure foreign rights in economic matters and international interest in the military — or, rather, in maintaining the non-military — status of the archipelago. Other maritime nations will be affected by the global reach of Soviet sea power, and Atlantic states will feel special concern about the performance of the Soviet Navy in and from Northern Waters. To grasp the full importance of recent and future maritime developments in the north, it will also be helpful to have some understanding of earlier events and of the broader international interest which is attached to the region.

THE PETER PRINCIPLE

It is worth remembering that Peter the Great was the first Russian ruler to recognise that maritime operations could be used as an instrument of state policy and to expand the reach of Russia's power and territory. It was he who gave orders for the Great Northern Expedition to explore the entire length of the Arctic coast, and he personally hired Vitus Bering to scout those distant waters of the North Pacific which still bear the name of that great Dane. Thus, the full continental reach of the empire was secured, with shores on Atlantic waters (in the Barents Sea and the Baltic) and coasts on the Pacific (in the Seas of Japan and Okhotsk, and the Bering Sea).

Bering's discovery in 1741 made Alaska a Russian province and was to be the empire's first, and so far its only overseas possession. This happened several years after the death (in 1725) of Peter the Great, and at that time the momentum of the imaginative tsar's maritime drive was already waning and Russia was settling into her tradition of a nation bound to the soil, with limited vision for the uses of the sea. The sale of Alaska to the US in 1867 may well be seen as a belated reaffirmation that Russia had resigned to the role of the landlubber among the great powers. In spite of some imaginative and masterly exploits (for example, Adm. Faddey Bellingshausen's circumnavigation of Antarctica at high latitudes, 1819-21) and in spite of the construction of a navy which was large in numbers and advanced in technical quality, Russia did not gain a true sense of the advantage and importance of using the seas.

One reason for this, of course, was that while the Russian Empire did succeed in its effort to reach the seashore in all directions (with the Indian Ocean as the only major exception), geographic, as well as political and economic conditions generally restricted vision towards the open sea. Russian coasts were either ice-bound during a good part of the year (as in the Arctic), or they were on enclosed seas with long passages to the open ocean and could easily be blocked by an enemy (as in the Black Sea), or both (as in the Baltic), as well as very far removed from the country's political, economic and population centres (as in the Bering Sea and the Sea of Okhotsk). Thus, nature, politics and tradition all favoured a careful and even fearful and protective attitude towards the sea. Actual experience from naval operations was often negative, and tended to reinforce the preference for withdrawn, defensive naval tactics and a passive maritime strategy.

The sale of Alaska took place shortly after the Crimean War (1854–6), when European maritime states (England and France) had demonstrated that their navies could be used to attack Russian territory, with raids in the Baltic, landing and support operations in the Black Sea and strikes even against distant Petropavlovsk on the Bering Sea. While not all of these actions were glorious successes — the Petropavlovsk campaign was inconclusive at best — they were nevertheless sufficient to force humiliating defeat upon Russia. The lack of stronger naval forces in the North Pacific quite obviously contributed (according to Adm. S.G. Gorshkov) to the Russian decision to let Alaska and the Aleutian Islands go at a reasonable price. The first, and so far the only major engagements of a Russian navy in major battle, during the war with Japan, also ended in miserable defeat (Port Arthur 1904 and Tsushima 1905), and in both the First and Second World Wars, the Russian Navy was used with utmost care and played a role which did not match its potential.

With this background, it is no surprise that present-day protagonists of Soviet sea power (including the present chief of the Red Navy, Fleet Admiral Vladimir Chernavin) prefer to look back to Peter the Great to establish historical roots for their mission and proof of their value. It is a measure of their success that they can claim now, as Adm. S.G. Gorshkov did some ten years ago, that in the Soviet Union's effort for 'the building of communism and the steady rise in the welfare of its builders, sea power emerges as one of the important factors for strengthening its economy, accelerating scientific and technical development and consolidating the economic, political, cultural and scientific links of the Soviet people with the peoples and countries friendly to it'.[1]

In little more than three decades, Soviet planners have made this a fundamental tenet of Russian policy and they have succeeded in securing the necessary resources — economic, technical and human — to turn traditional landlubber Russia into a modern maritime power with both military and commercial capability to operate with confidence in all waters. Russian vessels of all kinds and descriptions — with super-tankers as the only major exception in the merchant fleet and large aircraft carriers the only exception, so far, in the navy — now carry their red flag all around the world to demonstrate their presence and capability. Thus, 300 years after Tsar Peter's youthful experiments in shipbuilding and sailing on Lake Pleschev, fears arise that Russia has finally acquired its tool for extending influence and projecting power world-wide, against the interests of the traditional maritime powers.

The main threat is seen as coming from the Soviet Navy, especially from its submarines with their twin threat of nuclear strike against the United States (and any other target area) and enhanced ability to cut maritime supply lines, but the general strength of the Red Navy, with increasing range and capability for missions in distant waters, is also a matter of increasing concern.

To the possibility now arising that the Soviet Navy may soon be able to establish sea control in some critical areas in the event of crisis and war, must be added the danger that the Soviet Union may also be able to dominate important commercial trades. Here, it must be remembered that Soviet maritime strategy is much more than a blueprint for war at sea.[2] The Soviet concept of 'state sea power' is broad and multifaceted. To the threat it poses in war must be added the dangers of the challenge it presents in peacetime competition at sea.

SOVIET MARITIME STRATEGY

The spectacular growth of its navy provides the most remarkable evidence of the Soviet Union's emergence as a major maritime power but, as Admiral Gorshkov himself has pointed out,

> This in no way means that the sea power of the country depends only on the fighting strength of its navy. It must be regarded primarily as the capacity of the state to place all the resources and possibilities offered by the ocean at the service of man and make full use of them to develop the economy, the health of which finally determines all facets of the life of our country including its defence capability. In this context the concept of sea power to a certain degree is identified with the concept of the economic power of the state. Accordingly sea power may be regarded as a constituent part of economic power. Just as the latter determines military power, sea power, mediated by the economy of the state and exerting an influence on it, carries within it an economic and military principle.[3]

By any standard, the maritime development programme carried out by the Soviet Union in the last few decades (i.e. after the Cuban crisis in 1962) is a remarkable achievement. In a longer-term historical perspective, the most important element in the legacy left by

S.G. Gorshkov (and his colleagues, who helped the process) may well lie in the very concept of sea power as an all-inclusive instrument for the promotion of the national interest through maritime operations. In developing and arguing successfully for an integrated concept of Soviet state sea power, Gorshkov *et al.* broke free from the strictures of traditional Russian continental mentality and provided the nation with a new ideological rationale for a Soviet maritime strategy.

In this strategy, the navy is the most important single instrument, but commercial shipping, fishing and research are also significant elements wrapped into a joint economic and military principle with the promotion of Soviet state interest as its primary goal. In this broad sense, Soviet maritime strategy is defined by Gorshkov's dictum that 'Soviet sea power, merely a defensive arm in 1953, has [now] become the optimum means to defeat the imperialist enemy, and the most important element in the Soviet arsenal to prepare the way for a communist world'. Using the seas for transport, commerce and resource development, as well as an arena for competing with rivals and fighting enemies, is not an innovative achievement, but common practice for states with a maritime tradition. In a continental state with a landlocked tradition, on the other hand, the construction of a conceptual base for maritime expansion may be a major event with long-term effects for the state itself and significant implications for its relations with other states.

The Soviet concept of the sea power of the state is unique in unifying all maritime interests and activities as constituent elements of superior state interest. With the centralised system of direction of the national economy, with single-party control of the entire political process from decision-making to implementation, and with the ideological content of Soviet political principle and doctrine, this allows, and indeed commands, a maritime strategy in which all elements will operate for a common goal and purpose.

In the Soviet system, any and all activities which require any significant investment in equipment and personnel are controlled by the state. Thus, not only the navy, and research activities which require state subsidy, but commercial shipping and fishing as well, and maritime research, will be planned, developed and operated within a general framework of state interest. Here, the organisation of the economy, and the procedures for allocation of resources, add to the principle of state control and ensure that national political goals are built into maritime operations from the planning stage onward and used as a standard for the determination of priorities and for co-ordination of effort.

The difference of the Soviet system from the conduct of maritime operations in traditional maritime nations, notably in the commercial field but also largely in scientific research, is fundamental. There, non-military operations such as shipping and fishing will be organised in a structure of separate companies to be conducted in accordance with commercial principles of competition and profits. National policy goals for maritime activities are seldom fixed in any but the vaguest terms, and almost never will determine actual conduct of operations. In non-socialist countries, political principles, organisational structure and commercial practice combine to foster a fractionalised system and, moreover, with shipping and to some extent also fishing as industries in which economic liberalism and competitive individualism has dominated, the very idea of an overall national strategy is anathema to maritime thinking.

In economic terms, the Soviet system may be weakened by institutional rigidity and bureaucratic blundering, and proponents of the free market system will argue that Soviet maritime operations are inefficient. Nevertheless, the Soviets do compete with Western shipping, Soviet fishing does account for a very large share of total world catch and Soviet ocean research is carried on in all seas. Together with the increased strength and reach of the Soviet Navy, this adds up to a massive manifestation of maritime capability and sea power controlled by the Soviet state.

Reforms may be required to improve Soviet performance at sea, but the system has produced great results in a very short period of time. Indeed, if the economic shortcomings do exist, they may emphasise the determination of the Soviet state in its effort to establish and maintain a massive potential in all types of maritime operations, more than they expose a weakness in the system. To understand the full potential and to assess the broad implications of Soviet sea power, it is essential always to keep in mind that Soviet maritime strategy is cast in the mould of unified state purpose. This also means that in observing and analysing Soviet behaviour at sea, we must always expect to find political motives, general and particular, as causes that decide action.

ROLE OF NORTHERN WATERS

Geography and politics in combination give Northern Waters a special importance for the Soviet Union. When Tsar Peter was a young man, Archangel was the only Russian seaport and the one point where vessels could land and load goods in foreign trade. Peter's expansion towards the coast changed that by providing access to the Baltic (the Great Northern War, 1700–21), and he opened the way to the Black Sea (Azov victory, 1696). His initiative to the Great Northern Expedition shortly before his death, secured the expansion of the Russian Empire to the Pacific and laid the foundations for Russian control of the full length of the Eurasian coast on the frozen Arctic, from Archangel to Petropavlovsk (on Kamchatka). Subsequent territorial expansion, which has continued up to the present (including Soviet re-annexation of the Baltic states and annexation of parts of Finland and Rumania, as well as the northern Japanese islands), has consolidated Russia's grip on important coastal regions. Nevertheless, Russia/the Soviet Union has failed to establish warm-water ports with free and safe access to open oceans.

This failure to secure effective control of the passage to open waters from the Black Sea and from the Baltic (and the fact that far eastern ports which have a somewhat smaller problem in this regard are so far removed from the Soviet centre and from the Atlantic theatre) have left Soviet Russia with a problem that Tsar Peter had been the first to face. With all progress from those times and with all its maritime development, the Soviet Union still carries the essential characteristic of tsarist Russia as a continental power whose main coast is ice-bound and desolate and where the main opening to the sea is still in the far north-west, between the old port at Archangel and the Norwegian border.

While other ports suffer no serious disadvantage in peacetime and are booming with commercial operations, northern ports are essential for the present and future role of the Soviet Union as a global sea power with a large navy. From the Soviet point of view, the waters extending from the Norwegian mainland to Iceland and Greenland in the west, and from the Greenland–Iceland–UK line to Svalbard and the Barents Sea in the north, will be equally important as a gateway for the projection of Russian maritime power/Soviet state sea power abroad, as they are as a protective zone for its strategic submarines and a barrier against enemy seaborne threats against the homeland itself. In this latter context it should be remembered that Western powers interfered in the Russian revolution in 1918–19 by sending troops to the Murmansk–Archangel region (as well as to the far east). Defensive and offensive requirements point with equal urgency to a Soviet need for a policy to minimise enemy threat in and from Northern Waters, and to make them fully secure for Russian maritime operations. Secure protection and safe access for itself in Northern Waters simply are essential security requirements which the Soviet Union will always wish to meet.

The Northern Fleet, which operates out of bases in the Murmansk–Kola area and uses the Northern Waters as its primary training ground for defensive and offensive operations, is the largest of the four Soviet fleets. In sheer numbers, the Pacific Fleet is now larger, and growing, but the overall strength of the Northern Fleet remains greater, particularly in strategic and attack submarines, where its strike power still equals that of the other three (Baltic, Black Sea and Pacific) fleets combined. Numbers are imprecise measures of strength, but they suffice to illustrate the dramatic change from former times, when Russian naval strength was concentrated in the Baltic and the Black Sea, and Western fleets dominated the Atlantic, including the far north. With this increased military importance for the Soviet Union, political interest will be equally great.

The build-up of the Soviet Northern Fleet, and the gradual expansion of its area of training and exercise which has taken place in recent years, has altered the earlier maritime balance - with its western predominance in the north-east Atlantic - to a degree that very few, and probably not the Russians themselves, would have considered to be possible 25 years ago. For Europeans, for NATO, and, of course, for Norway and the Nordic countries, this presents a new threat, and a very clear and present danger in the event of war.

Even if a strategy is developed and measures are effectively applied to counter that threat, the old balance with Western predominance cannot be re-established in Northern Waters. What we have seen happening is a permanent shift where the only alternatives seem to be acceptance of Soviet predominance, at least in parts of these waters, or a considerable upgrading of NATO presence and operations to establish a new balance at much higher force levels and with improved readiness. The effects that this shift will have in the event of future war, for defence on the Northern Flank and for maritime control in the Atlantic and, with it, for the defence of Europe, could be momentous, but the political implications in peacetime could also be quite considerable, both by increasing general political awareness and concern, and by inviting and inciting moves to exert influence upon events and developments in the region.

The point that must be made is that not only the littoral states themselves, — namely, the small Nordic countries , Norway with Svalbard, Iceland, and Denmark with Greenland — will feel immediate concern for future military and political development in the north. With the strategic importance which control of these Northern Waters can have for the security of transatlantic sea lanes, as well as for the (sea-based) nuclear balance between the superpowers, political problems and military developments alike in the region will also affect all North Atlantic states and, of course, the Soviet Union itself, as the main perpetrator of the recent military change.

In this context, we must remember also that present interest and concern about Northern Waters has a prehistory, that earlier events have left a lasting imprint in the form of treaty provisions, and that prior experience from practice and operations, some of which have now been changed, will be conditioning factors in the evaluation of current events. Moreover, with the advent of offshore exploration for oil and natural gas and great expectations for future development in the Svalbard–Barents Sea area, many parties will look with particular interest for opportunities there. In so doing, they will lean upon history, at the same time as they will act under the impression of current developments and trends, with main emphasis on the role of the Soviet Union as a key actor in the region. To understand that role more fully, and to assess better the impact of the changes forced under the weight of Soviet sea power and superpower status, it may be helpful to recall some of the prior events.

ANTECEDENTS

Looking at present security problems and potential local threats from the Soviet Union in the far north, it is sometimes helpful to remember that the first fortress in Norway's northern coastal area was in fact built as early as 1307 on Vardoe Island. The present Vardoehus fortress is the third on the same location and was built over the period 1734–8 to stand sentinel against penetration from Russia. Norway and Russia have never been at war, but the historical monument that Vardoehus now is, is a proper reminder that suspicions of Russian intentions are as old as our awareness of their presence in the north. The quaint dolls-house quality of the old fortress belies its past importance, but the sombre grey radome and installations on the hill across the bay from the fort show that a guard begun more than 600 years ago must be kept even today.

In the days when the first fortress was built, the Northern Waters were a *Mare Norvegicum* in fact, and not merely in name. From the time when Ottar brought his amazing tale of travels to King Alfred's court in England (end of the ninth century), the Norse had dominated Northern Waters more or less at will for hundreds of years. However, the Norse fell into decline at the end of the fourteenth century, and present-day Norwegians need to remind themselves that it was not they, but the British and the Dutch and, to a lesser extent French Basques and Germans, who revived northern exploration.

In the sixteenth century, when maritime exploration opened the way for economic revival in Europe, progressive traders in England and the Netherlands wanted their share of trade and colonies in the Far East. The southern routes were controlled by the Spanish and the Portuguese, who had discovered them and, for a time, had the sea power, as well as papal blessings and a right under the legal precepts of the day, to defend their discoveries against intruders. The merchants of London and Amsterdam were therefore forced to send ships in search of a northern sea route over the top of the world to gain access to the fabled riches of Cathay (China).

The search began (with the Willoughby–Chancellor expedition in 1553) towards the north-east, past Norway and into the cold waters beyond, and soon proceeded into the north-west, between Greenland and North America. It took three centuries before a northern sea route was found (the Russian Great Northern Expedition had surveyed the coast and waters in sections, but Sweden's Adolf Erik Nordenskiöld was the first to sail the full length of the Northeast Passage in 1878–9, and Norway's Roald Amundsen found his way through the Northwest Passage over the period 1903–5). However, the unsuccessful early attempts did lead Willelm Barents to discover Spitsbergen (now Svalbard, on the basis of a reported Norse discovery in 1194, with no accurate location) in 1596, and when Henry Hudson came there in 1607, he reported on the rich occurrence of whale, walrus and seal in the region. This brought ships from many nations scurrying into these Northern Waters to exploit their riches and some, notably the Dutch and the British, had seasonal stations on the islands.

Thus, an effort which began as a strategic move to outmanoeuvre the Spanish fleet by finding a new sea route to the Far East, resulted in large-scale whaling, sealing and, later, fishing in Northern Waters with participation from several countries. With the growing maritime strength of the English and the Dutch, and the defeat of the Spanish Armada (1588), Spain soon lost its ability to control the southern routes. New principles of the freedom of the seas were developed (Hugo Grotius' *Mare Liberum*, 1609), and mariners and merchants could sail the seas at will, while the whalers and sealers and fishermen stayed on in the north. National claims for territorial sovereignty over Spitsbergen were not honoured and for centuries the islands and surrounding waters were used as a common *terra nullius/terra communis* by a variety of Europeans.

The English and the Dutch dominated these operations in the first period (the French did not stand up to the competition here, and moved to the American far north instead). Northerners played a minor role in early whaling, *inter alia* because they lacked capital for the investment in the large vessels required, and for setting up stations on land. Early in the eighteenth century, when whaling had ceased (due to depletion of the stock), groups of Russian trappers began to winter over on the islands (to hunt reindeer, polar bear and fox), and from the beginning of the nineteenth century, Norwegians began to dominate operations, primarily with their sealing in Spitsbergen waters, but also with fur trapping on the islands. However, Germans, Swedes and Danes also took part in the operations. Later in the century, a combination of scientific curiosity and spirit of adventure aroused fresh interest in the polar regions, and Spitsbergen, which was the most easily accessible part of the High North, attracted attention and visitors from many countries. The introduction of the steam engine greatly facilitated operations and, more importantly, made distant water trawling in the north a major activity for European fishermen.

Apart from trappers, who stayed the winter on the islands, all activities at Spitsbergen were confined to the summer season. A request for financial support to establish a permanent 'colony' on Spitsbergen was rejected by the Norwegian government, and an attempt to gain international recognition of territorial sovereignty for Norway over the islands (1870-1) failed when Russia objected. Thus, no regulatory authority was recognised and no system to secure law and order in the activities and operations existed. At the beginning of the present century, it became obvious that the lack of all law could not continue, and that some jurisdiction must be established.

THE SPITSBERGEN TREATY

Coal-mining made it necessary to establish a legal regime for Spitsbergen. The existence of coal on the islands had always been known (coal was used for cooking whale oil in the seventeenth century), but in 1899, when a Norwegian sealer brought back a full cargo to compensate for a poor catch, plans for commercial mining were promoted and a number of more or less speculative enterprises were set up. Initiative often came from Norwegians, and Norway provided most of the labour force, but capital frequently came from foreign sources. Companies for Spitsbergen mining were organised in several countries, including the United States. Thus, a need arose to regulate title to land and to provide security for labour and capital that had not existed in those hundreds of years when only the living resources in the sea and on land were exploited.

Norway was the country most immediately concerned and after heated debate in the Norwegian and the Swedish press, the government began soundings (in 1907) amongst interested parties, and in 1914 a conference was called in Oslo (then Kristiania) to consider a proposal worked out by Norway, Sweden and Russia as the three countries with the greatest interest. In addition, Denmark, France, Germany, Great Britain and the Netherlands, all with companies engaged on Spitsbergen, took part. Under the proposal, Norway, Sweden and Russia would form a tripartite commission to manage the islands as an international area (*terra nullius*) and to organise a minimum administration for mining and other exploitative activities. The Americans were most critical. While accepting the tripartite structure, Washington submitted a separate proposal (prepared by Robert Lansing) that would, *inter alia*, give all signatories a right to veto decisions made by the commission.

On the surface, disagreement on the 1914 proposal was mainly over administrative arrangements (and taxation), but underneath there was deep concern in some capitals about underlying political issues that were not yet clearly defined. Robert Lansing, the American international lawyer who advised the US State Department and was a main cause for the rejection of the tripartite proposal, helped define those concerns in an article which was written in 1914–15 ('A unique international problem') and published in 1917 (*American Journal of International Law*). Here, he argued that a regime for Spitsbergen must be based on territorial sovereignty, and suggested that sovereignty be vested in 'a neutral Scandinavian power'. His opinion was important, for in 1915 Lansing was appointed Secretary of State. In that capacity he proposed in a memorandum to the President (21 September 1918) on the peace settlement in Europe, that 'The sovereignty of the archipelago of Spitsbergen [to] be granted to Norway'.

One should be careful not to overestimate the clairvoyance of Lansing and others in these matters, but his scepticism of a troika system in which the great power Russia might dominate over two small partner states without any corrective vote for other participants, and his suggestion that legal authority on the islands must be based on sovereignty vested in a small and neutral country, indicate a clear understanding of the larger political issues. His proposal that Norway's sovereignty over Spitsbergen should be recognised as part of the peace settlement was more than a move to clear up a political tangle over mining rights. With hindsight, we may see in the proposal a clear determination to prevent future great power involvement on the islands, and to use small power neutrality to prevent military conflicts on, over or from Spitsbergen.

Norway was eager to acquire sovereignty and willing to do so on terms that were acceptable to the victorious great powers. To them, the Treaty that was worked out and adopted in 1920 as a separate agreement, but in connection with the peace negotiations after the First World War, must have seemed to be the best of all possible solutions. The Treaty recognised Norwegian sovereignty over Spitsbergen, but on conditions that gave citizens and companies from other countries equal rights of access and equal treatment in exploration and exploitation of all natural resources and in other commercial activities. Furthermore, prospecting for and extraction of mineral resources was regulated by a separate Mining Code which was approved by the parties before the Treaty entered into force, and the Treaty and the Code put specific and strict restrictions on taxation and other dues.

Moreover, and this was most important in a political and strategic context, Norway was to make sure that no naval bases or fortifications were established on the islands, 'which may never be used for warlike purposes' (Article 9). Again, we should be careful not to overestimate the ability of diplomats at the end of the First World War to analyse long-term tendencies in international power politics and trends in military technology and war-fighting techniques. Nevertheless, this neutralisation in practice of Spitsbergen did prevent future competition for control of the islands with a view to using them as a base for military action. Thus, a potential source of conflict was removed from Europe's political map, and one of the world's first and most effective arms control measures was adopted.

Pre-war expectations for mining profits on Spitsbergen were quickly frustrated and during and after the war there was a sharp reduction of activity. Since the 1930s, only Norwegian and Russian companies have been active. Today, there are three mining towns in operation, the Norwegian Longyearbyen (named after its former owner, the American John M. Longyear, who sold his Arctic Coal Company to a Norwegian consortium in 1915) and the two Russian towns of Pyramiden and Barentsburg (operated by Arktikugol). Productivity in the Norwegian mine is high by European standards, but operating cost is higher and the company now runs on severe deficit. The company (Store Norske Spitsbergen Kulkompani) has been taken over by the government (with less than one in 1,000 shares remaining in private possession) and is heavily subsidised. The Russians, with a population twice that of Norway, have a smaller coal production.

It would seem that political rather than economic motives are the main reason for maintaining production and population at present levels. In granting funds for continued mining, the Norwegian Parliament has made it clear that this is necessary as a manifestation of sovereignty. It seems reasonable to assume that the Russians, while needing coal for use on the Kola Peninsula, are also willing to pay the high cost of maintaining a population ratio of two Russians to each Norwegian because this gives them better opportunity to follow and inspect activities on the islands to make sure that Treaty provisions are scrupulously observed. It should be noted, though, that the fact that Russians are able to make sure that no one else violates the military restrictions also makes it impossible for them to press false charges of illegal operations. The sum total is that their presence and freedom of movement contribute to securing the main purposes and principles of the Treaty.

SOVIET DESIRES

The Spitsbergen development after 1914 ran counter to Russian desires. In its separate peace settlement with Germany (Brest–Litovsk Agreement, 3 March 1918), the Soviet government was forced to accept an obligation to support equal status for Germany and Russia in any future international arrangement for the islands. The revolution, and the conclusion of a separate peace treaty with Germany, blocked Russian participation in the peace negotiations in Paris, including the negotiation of the Spitsbergen Treaty. Any hope that Russia might have had for a special and preferred position on Spitsbergen was lost, and the obvious dislike for the new Soviet regime amongst the victorious powers that was manifest in their intervention against the Red forces in the Russian revolution, clearly favoured a separate small-state solution for the administration of the islands.

In principle, one might have expected Russian satisfaction with the 'Norwegian solution' for Spitsbergen. After all, it would prevent any future use of the islands as a base for attack against the Soviet Union: one main reason why some powers (France, in particular) favoured an agreement for Norwegian sovereignty with strict obligations to prevent naval bases and fortifications and to make sure that the islands were never to be used for warlike purposes, was determination to prevent Germany from gaining any foothold in the Arctic. By withdrawing Spitsbergen from great power influence, the Treaty would generally prevent foreign designs on Russia in the Arctic, and preclude in particular German plans to exploit Spitsbergen as a base for future naval operations in Northern Waters. Moreover, Russians certainly could benefit quite as much as others from the equal treatment clause in commercial operations.

The Soviets, however, were not satisfied. Norway wanted Russian acceptance of the Treaty before it entered into force, but this proved difficult. When negotiation of the Spitsbergen Treaty was completed, the Soviet government informed Norway (in February and in May 1920 and, again, in January 1923) that it would not consider itself legally bound by any solution where it had not taken part in the negotiations. Then, a year later (in February 1924) the Soviets declared that Russia would be willing to recognise Norway's sovereignty over Svalbard provided that Norway gave full (*de jure*) recognition of the Soviet government, and a joint declaration to this effect was issued.

Of course, it is difficult to know if the Soviet government feigned reluctance to accept the Treaty in order to win recognition for itself, or if recognition from Norway was so important that it was willing to forget its opposition to the Treaty in order to get it. In any event, Moscow did not accede to the Treaty before 1935, ten years after it had entered into force and after a new request for acceptance of the Treaty had been made. At that time, accession had become necessary for Russian mining operations on Svalbard, as the islands were now called.

In the meantime, Moscow had promulgated the sector declaration (1926), which claimed Soviet sovereignty over all lands and islands, unknown as well as known, between the Soviet north coast and the North Pole. The motive clearly was a desire to secure the northern region against foreign intrusion, which had been seen in recent Canadian and US moves on Wrangel Island (off the East Siberian coast) and in Norwegian dominance in sealing expeditions in the Franz Josef Land area (to the north-east of Svalbard).

Experience during the Second World War, with the German occupation of Norway, the sea battles in Northern Waters and, above all, the severe losses imposed upon the Murmansk convoys by German air and sea attacks, convinced the Soviet government that an effort must be made to improve the Soviet position in Northern Waters. To this end, a demand was put to the Norwegian exile government in November 1944 to cede Bear Island (in the middle of the Barents Sea, and legally a part of Svalbard) to the Soviet Union and to accept revision of the Spitsbergen Treaty to allow joint Soviet–Norwegian rule (*condominium*) on the other Svalbard islands. Furthermore, the military restrictions in the Treaty should be abolished to give the Soviet

Union and Norway an equal right to have troops and fortifications on Svalbard, and a joint obligation to defend the islands.

These demands to a wartime ally before the war against the common enemy Germany had been concluded were justified by Foreign Minister Molotov as being necessary to secure the passage from Soviet northern ports to the Atlantic. He argued that passage was blocked in the Bosporus–Dardanelles (demands were put to Turkey at the same time to secure that passage) and in the Baltic approaches (where Bornholm was liberated by the Russians at the very end of the war and then occupied for more than a year before the troops left) and he stated flatly that the Soviet Union was a great power that could no longer tolerate a situation whereby it was excluded from free access to the world oceans. The demands were repeated in 1946, when earlier declarations from the Norwegian exile government that it was prepared to make major concessions had lapsed, but have since been dropped, apparently because revision of the Treaty would require consent from the other parties, and such agreement would not be forthcoming.

It seems fair to assume that the demands against Svalbard in 1944–6 may be seen as an optimum aim for the Soviet Union at the end of the war, as well as an indication that Moscow will use a momentary advantage against its small neighbour to promote its own interests in the north: the demands were first put at a time when Soviet troops had begun the liberation of the Norwegian border area and Norway's exile government was afraid that refusal to negotiate could lead to further demands, or to Soviet refusal to withdraw when the war was over.

It also seems fair to conclude that further pressure for concessions was discarded because other parties protested strongly when the Soviet demands became known to the public (in January 1947). At that time, the euphoric mix of admiration and fear towards the Soviet Union was beginning to give way to more realistic attitudes and the Norwegians, and the Russians themselves, were aware that the spirit of compromise and concession that had dominated Western attitudes towards the Soviet Union for a period at the end of the war was now disappearing (namely, the post-Potsdam reactions in contrast to the Yalta 'give-away').

A most important conclusion, then, will be that the fact that other states, including all the West European countries and the United States who are now Norway's allies and partners, are party to the Spitsbergen Treaty with clear interests, political and economic as well as military, in preserving the Treaty regime, will act as a powerful brake upon Soviet efforts to gain a position on the islands and an influence in the area which the Treaty aimed at preventing. In this respect, the Treaty has a new and important function in providing an international balance against Soviet pressure.

Under the circumstances, the Soviet Union has adjusted to living with the Treaty and to maintaining presence and activities on Svalbard at a relatively high level. Coexistence between Norwegians and Russians on Svalbard has seldom caused any serious difficulty, but it is perhaps characteristic that efforts by the Norwegian government to 'normalise' Svalbard through legislation and more effective administration have generally met with a degree of passive resistance, and that the Soviets have regularly claimed special status among the parties to the Treaty. Pointing to the fact that together with Norway, the Soviet Union is the only party with permanent presence, demands are regularly made for prior consultation before new laws and regulations are adopted.

Here, the Soviet government deliberately disregards the fact that under the Treaty it is not the Soviet state, but only Soviet 'nationals' (i.e. persons and companies) who are entitled to be present and to engage in activities on the islands. Another characteristic feature of the Svalbard relationship is that the Soviets have frequently complained that Norway is acting contrary to the Treaty (e.g. by having military aircraft and vessels visit the islands). In sum, it is apparent that the Soviet Union wishes to maintain maximum freedom for Russian activities on Svalbard and to keep maximum 'guard' against all others (for example, frequent 'inspection tours' with their own helicopters and 'friendly visits' whenever a new group sets up camp or opens new activities). Norwegian media sometimes charge 'suspicious behaviour' on the part of the Russians that could involve various forms of military preparation (for example, the helicopter base) and flagrant disobedience of regulations. This may be more indicative of the special 'Svalbard atmosphere' than of real threats, but it does reflect mutual suspicion.

MUTUAL ACCOMMODATION

With these prevailing suspicions, coexistence in the north depends on mutual accommodation. Norway and Russia have sharply different interests in the region, but an apparent common interest is avoiding confrontation which could escalate quickly from a local dispute to a major international conflagration. Thus, Norway, finding security in NATO, has taken care to avoid measures that could be seen as posing an offensive threat against Soviet territory. These self-imposed limitations, which include refusal to establish Allied bases in Norway or to accept nuclear weapons on Norwegian territory — as long as the country is not attacked or threatened with attack — and restrictions upon exercises near Soviet territory, serve a political purpose as well, by encouraging the Soviet Union to abstain from military and political moves that could be seen as a threat against Norway and call for reconsideration of the defence limitations.

It is generally accepted that the extreme concentration of military strike power, conventional as well as nuclear, in the Kola region is aimed at targets beyond Norway and that an attack on Norway will only take place in connection with (or lead to) major hostilities between the Soviet Union and NATO. While there is no doubt that Soviet capability for an attack on Norway has increased substantially in recent years (and therefore led to prestocking of Allied arms and equipment in Norway) and Soviet exercises (for example, amphibious training) show an ominous pattern, it is still felt that the Soviet Union is also careful to avoid measures that will increase tension and risks of open confrontation. Episodes do occur, notably in near waters and offshore areas, but in such cases firm Norwegian reaction has proved effective, and a tacit understanding about mutually acceptable behaviour seems to have emerged.

At the same time, there is no doubt that recent developments make for a more difficult situation, for defence, and in political terms as well. Parallel with the very rapid growth of the Soviet Navy and Soviet maritime interests, a new international Law of the Sea has developed, with coastal state jurisdiction being extended to include continental shelves and maritime economic zones. In the north, the entire area from the Norwegian mainland and northward past Svalbard up to approximately 84°N is now subject to national jurisdiction for resource development and environmental regulation under the terms of applicable international law. This means that the entire maritime access to/from Soviet northern ports will be through waters subject to neighbour state jurisdiction for those purposes.

The new legal status of the northern maritime area will have very significant implications by setting new rules of behaviour for many of the activities and operations in Northern Waters, and by facilitating entirely new operations, such as offshore development of hydrocarbons. In principle, there is no linkage between national maritime resource jurisdiction, and navigation and military activity beyond territorial limits. Nevertheless, a new *modus vivendi* must be established. The parties will have to adjust themselves to a new situation with other juridical and practical conditions, and a degree of accommodation will be required to prevent unnecessary tension and episodes.

As of today, every indication is that Northern Waters, including the Barents Sea area between the mainland and Svalbard, from the continental edge (at approximately 15°E) towards Novaya Zemlya, will be every bit as important for its resources as it has ever been. Still rich in living resources and an important source for the marine protein supply, the area may soon become even more important for its hydrocarbons. Indeed, some companies regard the Barents Sea continental shelf as one of the most promising, undeveloped petroleum provinces in the world today. This adds greatly to the importance of jurisdictional problems which remain to be settled, including the border problem between Norway and the Soviet Union and questions about rights in the Svalbard area, and to political questions in resource development.

Norway and the Soviet Union are the only two countries with coasts on these important waters. This means that it is up to them to negotiate the borders between their respective areas of jurisdiction and to make the necessary arrangements for regulation of common stock of living resources and, if and when necessary, to reach agreement on practical arrangements in offshore development, for example, for security and support services, environment standards, etc. In short, with the new legal regime and with increased need to regulate resource operations and the environment, bilateral relations between Norway and the Soviet Union will increase in intensity and in complexity, with a need on both sides to develop a policy for this purpose.

One effect of these developments is that the traditional, multinational pattern which dominated operations in Northern Waters for several centuries, and where Russia had a relatively minor role, has been suddenly and completely changed. As a result, the participation of West Europeans has sharply decreased and an important economic incentive for their continued interest in the north has disappeared, while Norway and Russia, as coastal states, and the Soviet Union with its navy, have far greater interests than before.

It is difficult to foretell the effects that reduced presence of other nationalities may have upon future relations in the far north. As of now, the British and the French, who used to be very active in northern fisheries, have practically disappeared from Northern Waters (while others with no tradition, such as the Spanish, now come in great numbers to exploit fish resources under the non-discriminatory regulations in the Svalbard zone). And in defence, the United States is now more immediately concerned than the Europeans with the latter's reduced naval capability.

Offshore petroleum activities, if and when developed on a large scale, may attract wide participation, particularly from Norway's close allies and partners, who have the capability as well as the need for petroleum development. Such participation, however, will be by permission and not by right, for the coastal state has sovereign and exclusive rights to resources on the continental shelf, with full authority to control their development.

In these questions, littoral Norway and outside parties will have different economic and political interests. In sorting them out, as governments and companies must do, thought must also be given to the wider implications of a situation in the north where Norway and her allies will have competing interests in political and economic affairs, while common interests prevail in military affairs and security. The very possibility that popular thinking, and public opinion, will be affected in the long run by prevalent concerns and interests, and that disharmony between allies in important issues of energy and resource development may affect opinion and behaviour in other matters, including vital questions about security and defence, requires sober thinking and accommodation to new realities.

CURRENT PROBLEMS

At present, the single most important unsettled issue in the bilateral relationship between Norway and the Soviet Union is the lack of a fixed maritime border. Beyond the 12-mile point from the mainland, they have a 1,700 km-long common border that remains to be settled. Negotiations (or rather, talks) have been going on, at irregular intervals, for 15 years or more without visible result. A very large area (155,000 km^2, or more than the Norwegian part of the North Sea) is disputed. The area could hold considerable petroleum resources and is rich in marine living resources. Thus, great economic value is at stake.

At the same time, this middle part of the Barents Sea is of high strategic importance. Thus the full spectre of Soviet maritime interest is involved. Furthermore, the Soviet claim for a border that would follow the sector line rather than the usual median line does involve a degree of national prestige, as well as a question of principle, where acceptance of the sector line could eventually serve to promote ideas about more extensive national jurisdiction than usual within the maritime sector.

Strictly speaking, the border line question is a single issue and should be resolved on its own merits, on the basis of relevant criteria. Little is known about the negotiations themselves, but Soviet sources have hinted in various ways that a border solution would be easier if it could be seen in broader perspective. Carefully avoiding direct linkage with other problems and proposals for package deals, these hints suggest advantage from a wider understanding and emphasise common interests in the north as a cause for a special relationship between Norway and the Soviet Union.

On this basis, it would seem fair to assume that the Soviet failure to be more forthcoming in negotiations, and to state a will to compromise (as Norway has done), could signal something quite different from an uncompromising attitude: moves to a broader understanding could be exactly what they appear to be, and reflect an honest self-interest to solve other problems which the Russians see as being connected with the border problem.

The border between Norway and the Soviet Union in the Barents Sea is a border between two coastal states, and a divide between their respective areas of jurisdiction in regard to resource exploration and other rights defined in international law. As already noted above, coastal state offshore jurisdiction does not affect or interfere with the freedom of the sea in other regards, and particularly not with military operations. Nevertheless, the Barents Sea line will divide between two distinct and different economic and political systems and, as long as Norway is a member of NATO, between two different military alliances.

Relations across that border in that particular area are of paramount importance to Norway and the Soviet Union as the two states directly involved, but the management of those relations could be an important element in general East–West relations as well. If bilateral relations could be exploited to promote a special, regional relationship and particularly, if that relationship could be utilised to alienate outside parties from presence and influence, the Soviet Union could hope to gain greater influence for itself. Such thinking may seem far-fetched to the Western mind, but it could come naturally to a mind trained to the concept of state sea power and an overall maritime strategy.

For a small country such as Norway, there can be special risk and hazard in being left alone together with a superpower such as the Soviet Union in an area of strategic importance where difficult and sensitive problems must be settled through bilateral negotiations. Typically, voices have been heard in Norway arguing that it is better to accept Soviet demands for a sector line border now, than to risk more excessive demands later. Others have suggested that a better solution could be 'bought' for the border by offering a package with concessions in other questions — for example, in relation to Svalbard.

Spokesmen for such attitudes seem to disregard the wider implications and to forget, as did the Norwegian exile government when it was prepared to yield to Soviet demands upon Svalbard in 1944, that other parties may, after all, have an interest and feel a concern that will make it difficult for the Soviet Union to force its demands. They also overlook the fact that yielding once to an excessive claim or a tempting deal may invite further demands on another occasion. What they do recognise, though, is that bilateral relations with Russia are difficult in the north, and that Soviet demands may be expected to be greater and different from those to be faced in similar negotiations in other areas.

The risk — or advantage — of linking bilateral border negotiations to questions about Svalbard can but underscore the complexity of problems and issues in Northern Waters and indicate the care which the parties must show in dealing with them. The risk of superpower domination, or attempts to dominate, is quite obvious, yet the Soviet Union knows that it acts in an area in which other parties see strategic interests of their own, and where they find a legal base for their concern. Here, the Svalbard continental shelf problem becomes important.

THE SVALBARD SHELF

The Svalbard region includes approximately 50 per cent of the Norwegian offshore area north of 70°N. It is rich in marine living resources, and may have a significant hydrocarbon potential. The second and probably the most difficult legal issue in the north is the question about the status of the Svalbard zone. While there can be no doubt that Norway, with sovereignty over the islands, will have jurisdiction for regulation of resource exploitation in the maritime zone, questions arise about the conditions for operations there. If the provisions of the Svalbard Treaty apply, nationals from other countries are entitled to equal access and rights of exploitation on favourable conditions. If Treaty limitations upon Norway's sovereign rights in regard to resources and commercial operations cannot be interpreted to apply beyond the territorial limit, the general principle of exclusive coastal state rights will apply. The economic value involved could be great, and the difference in participation and development policy would be very considerable.

The Norwegian position is that Treaty restrictions upon sovereign rights cannot be interpreted to extend into the offshore area. Other parties have protested against this, or they have reserved their position, which could mean that they may want to discuss or negotiate the issue. To prevent premature conflict, Norway has established a special fisheries protection zone with non-discriminatory fisheries regulations for the Svalbard zone. For minerals/petroleum exploration, however, the usual principles and regulations apply, but actual exploration with an intent to develop resources so far has not been permitted. The wish to avoid premature conflict is obvious here too, and other parties have abstained from moves that would press the issue. Mutual restraint, therefore, is entertained on all sides.

The Svalbard shelf disagreement is interesting from almost any point of view, and calls for special consideration for its role in relation to the balance between national, bilateral (Norway–Soviet Union), and international interests in the region. Characteristically, the first reservations from out-of-area states (the UK and the US) came at just the time when the first round of border negotiations were to begin (in 1974, after earlier preliminary exchanges) between Norway and the Soviet Union. A very long section of that border will be between the Svalbard Islands and Soviet islands to the east (Franz Josef Land and Novaya Zemlya), with nearly 100,000 km^2 in dispute. There is no need to go into details here, but one point bearing directly upon the wider, international aspects of relations in the north, and upon the negotiating position *vis-à-vis* the Soviet Union should be mentioned.

While maritime border negotiations are a bilateral concern for the littoral states involved, states whose nationals have a legal right of equal treatment, as in the Svalbard case, can also claim an interest in protecting, and indeed, a right to protect, the rights of their nationals. Hence, by reserving their position on the question as to whether Treaty provisions shall apply on the continental shelf or not, other parties do in effect point out that they may have interests and rights which the negotiating parties will have an obligation to take into account.

Stated in more direct terms, reservations *vis-à-vis* Norway on the shelf issue may serve as a double warning: firstly, for Norway to remember that the rights/interests of other parties may be involved and, consequently, that Norway cannot make excessive or unwarranted concessions to the Russians without encroaching upon the rights of others; and secondly, the reservations served notice to the Russians that any pressure upon Norway for excessive concessions or special arrangements relating to the Svalbard zone could be seen as an infringement on the Treaty regime. In this context, the parallel with the affair of the Soviet claims upon Svalbard, 1944–6, and the stabilising effect of the wider international interest through 'membership' in the Treaty, will be quite apparent.

In this context also, the Russians must be aware of the obvious paradox in the Soviet position: in some respects, Russia will emphasise the bilateral quality of relations in Northern Waters, and the need for and advantages of a special relationship (former Minister for Fisheries Ishkov liked to talk about the Barents Sea as a 'joint Soviet–Norwegian Sea'). Nevertheless, the Soviet Union has formally protested against the Norwegian position on the shelf issue and, by thus taking a position that would extend the full Treaty regime to the shelf, is pursuing a line that would lead to free and equal participation for all parties (i.e. their 'nationals') with broad international involvement in the Svalbard zone.

The Soviet protest could be a blunder, or merely be ill-considered, but it could also open the way for political tactics, with withdrawal of the protest and support for the Norwegian position at an opportune moment. Here again, the Russian propensity for broad and inclusive thinking in questions of maritime strategy and about the nature, quality and aims of state sea power could be important for the future development. On the other hand, playing around too much with arguments and positions in important and complex Northern Waters problems could prove counterproductive. In this regard, the plans, policies and production programmes for offshore petroleum production could add further complications.

OIL ON TROUBLED WATERS?

In addition to the evolving strategic interest and the effects of the Law of the Sea, prospects for active petroleum development stand out as a major force for change in Northern Waters. Seismic surveys and exploratory drilling have been carried out on both sides of the still unsettled border, so far without commercial finds being made, but expectations remain high and oil companies and politicians alike appear convinced that the Barents Sea will prove to be an important future oil province.

At an earlier stage, some concern could be noticed on the Soviet side about the possible effects of offshore operations, and the potential use of installations for military purposes. For this reason, some uneasiness could be noticed in Norway, with fears that inclusion of foreign, i.e. Western oil companies on the northern shelf, might cause Soviet reactions. Nevertheless, international companies were included in the first concessions off northern Norway, and now act as operators on several blocks. No reaction has been noticed from the Soviets — indeed, they have signalled clear interest in having foreign companies participate in joint ventures on the Soviet side, but so far without visible result.

Even so, the decision not to exclude foreign participation from the first round of concessions on the Norwegian side was an important decision with quite considerable political implications for future relations in the region. Exclusion in the first round could have left an impression of being an additional self-imposed restraint, but this time of a non-military nature, to avoid possible trouble with Russia. As such, early exclusion could have become a precedent that could not easily have been changed and that could have invited demands for additional restraints on 'alien participation' to reserve the northern shelf as an exclusive preserve for littoral states alone.

In avoiding that trap, Norwegian authorities have in fact made sure that an international presence may be maintained in the north, much in the pattern of earlier practice, but in radically new circumstances. An added bonus, from the Norwegian point of view, could be that international participation on the Norwegian mainland shelf may serve to counter demands for access to the Svalbard shelf on Treaty terms. Furthermore, by their presence, and by tying European and American energy interests to these Northern Waters, international companies may in effect give additional insurance against possible Soviet pressure for concessions and influence in the area. In this sense, there could be a non-military security guarantee for Norway in future petroleum development in the north.

The above remarks may seem skittish and timorous for fear of Soviet intentions. There is no hiding the fact that the writer is sceptical of Russian wants and desires for securing the north as much as possible for Soviet interests, and that he sees a need to act with care and determination in dealings with Norway's superpower neighbour. History, and concepts and goals elaborated by responsible Soviet spokesmen, *inter alia* for the development and use of Soviet state sea power, are sufficient proof that security must be guarded in Russia's near areas. On the other hand, the success that Norway has had, in spite of some risky lapses (namely, the first response to the 1944 Svalbard demands) is such that living next door to Russia is quite tolerable and could offer interesting opportunities for mutual advantage. In this context, co-operation in various aspects of offshore petroleum development in the north could be particularly promising.

The Soviet Union is the largest oil producer in the world, and the largest proportion of that production now comes from northern resources (West Siberia). The very rapidly increasing gas production is also concentrated in the Siberian north. While natural gas resources are enormous, known oil resources are being depleted and new finds are necessary to maintain production at the present level. It is assumed that the extremely wide continental shelf in the north will hold very large reserves, particularly in the Kara Sea and presumably also in the Barents Sea. For obvious reasons (ice and climate, distance from support bases and facilities, transport), development will begin first in the Barents Sea, before moving into the vast unexplored shelf areas further to the east.

Base facilities are being built near Murmansk, and personnel are trained there. However, the Soviet Union suffers from a lack of offshore technology, and will depend on Western technology for early development off its own coast. Exploratory drilling has proceeded in some locations in the Barents Sea, with equipment bought from the West. As far as we know, no commercial find has yet been made and progress has been slow, apparently due to lack of experience and qualified personnel. However, if a major find of oil is made (there is no need for new gas resources), rapid speed-up of exploration and development is expected. Soviet talks with a number of Western companies, including a Norwegian consortium (BOCONOR) must be seen in this light, as a tentative opening for rapid development if a major find is made.

Norway would seem to be an obvious and natural partner for offshore development in the Soviet Union. It has an excellent record for offshore expertise and technology and, equally important, has ports and could provide offshore bases and support facilities for operations on the Soviet side, to relieve the congested Murmansk region. Moreover, Norway is exploring its own northern shelf, and co-operation could benefit both parties. Further still, earlier Soviet suspicions about Norway's offshore operations have subsided and danger signals from earlier episodes have disappeared.

In one episode (in 1985), a Soviet frigate deliberately cut the cable from a Norwegian seismic vessel operating well inside the Norwegian mainland zone and in an earlier episode (1983), a Soviet drill ship spudded a well exactly on the median line, thus seemingly provoking Norwegian opinion by moving on to the disputed area. The Norwegian government chose to disregard the element of provocation that might have been intended, and said instead that the Soviet drill ship operated on the safe side of a margin of error about the location; but, it was emphasised, if Soviet drilling proceeded further to the west, into the disputed area, that would indeed be a provocation, and Norway would react accordingly. For a time, Russians were eagerly seeking information about potential reactions, without getting definitive replies.

It is impossible to know whether the episode was a deliberate testing of Norwegian responses and nerves, or if it was a mere blunder, but the end result was that a firm Norwegian position was marked, and that the Russians chose to respect it. In other words, a *modus vivendi* has been established, whereby the median line has been defined as the western limit for Soviet exploration. By parallel definition, the sector line must be seen as an eastern limit for Norwegian exploration, and the area between as a protected zone.

In the other episode, where the cutting of the cable appeared to be quite deliberate and particularly ugly, a very firm protest from Norway resulted in an apology from the Russians, who also agreed to pay compensation for the lost cable. An interesting feature here was that the Soviet frigate may have been helping a Soviet submarine operating nearby, possibly calibrating instruments. In any event, the result was an explicit admission of fault — accidental, according to the Russians — and acceptance of the fact, in practice and in principle, that naval activities shall not interfere with petroleum operations on the continental shelf under Norwegian jurisdiction.

Thus, both episodes served to clear away doubts and to define rules of behaviour in questions about respective areas of jurisdiction, and about the relationship between minerals development and military activity in the economic zone. In this way, they have definitely stabilised the situation and contributed to improved relations in Northern Waters.

Moving into direct co-operation, however, is a difficult and complex problem. In response to a series of Soviet soundings, the Norwegian government has stated approval of technical co-operation in offshore development, and laid down some ground rules. Thus, co-operation may proceed on normal business terms, within established rules (including COCOM restrictions and OECD credit rules), and without requiring or meriting government agreement as a political umbrella. Only petroleum prospecting itself can show if such co-operation will actually take place, and negotiations are in the hands of the companies.

At the same time, it must be clear that if and when major development does occur, agreement must be established on a range of practical and technical matters, including safety standards, liability for damages and pollution, search and rescue, etc. Here, models exist in the North Sea and other areas, and should not cause serious problems. Indeed, such co-operation should help the development of normal relations with the Soviet Union in maritime matters generally, and not only in Northern Waters.

Nevertheless, while the Norwegian government is firm in its position that technical co-operation with the Soviet Union should be just that — i.e. technical and not political, practical and not ideological — some very important policy choices will have to be made that will influence the political development and relations in the northern region and may play a role in general East–West relations. Here, choice of partners, and terms of co-operation could be decisive.

Co-operation in the north could be bilateral, between Norway and the Soviet Union as the only littoral states in the area. It could also be more openly international, and continue the tradition of broad participation in activities in the European north. The undeniable technical expertise and broad experience of many countries would be as valuable here as they have proved to be in the Norwegian offshore development up to the present. The political implications could be important, and lend support to the prevalence of the essential political principles that were built into the Svalbard regime. Moreover, a generally open attitude to international presence and participation could help reduce tension and pressure in some issues, including those relating to Svalbard. For Norway in particular, and for relations with Western countries, this could be an advantage.

Mere two-sided co-operation between Norway and the Soviet Union would be less attractive, and pose obvious dangers by eventually making Norway more dependent upon its superpower neighbour and, as a result, more easily exposed to pressure. An alternative would be to include other Nordic countries, to establish a pattern for 'Nordic Cap' co-operation (i.e. between the European countries with territories north of the Arctic Circle). The Scandinavian countries (i.e. Norway, Sweden and Finland) follow that pattern between themselves in a number of fields, but doubts arise about including the Soviet Union.

Co-operation in offshore development is a more specialised matter, and Finland, as well as Sweden, who do not have a coast in the north, is eager to benefit from it. They both have technical and financial resources, but need a physical base, and political as well as legal sanction for a role in the offshore game. Soviet sources seem to favour such a role for Finland and for Sweden. It goes almost without saying that a system of exclusive Nordic–Soviet co-operation will meet with objections, particularly in Norway and among her allies and partners. By emphasising its regional quality, it would mark distance to outside parties, and it would separate Western strategic interests from other developments. It would also promote the traditional Russian preference for regional arrangements, with their possibilities for great power dominance.

In sum total, developments in the north offer fascinating vistas and intriguing challenges. Here, within a small geographical area, we shall be able to observe the Soviet approach to its own role as a major maritime power in most of its aspects, and to test our own responses to it.

NOTES

1. S. G. Gorshkov, *The sea power of the state* (Pergamon Press, Oxford, 1979).
2. Finn Sollie, *Soviet sea power and relations in Northern Waters — trends and goals in Russian maritime strategy* (Norwegian Atlantic Committee, Oslo, 1985).
3. Gorshkov, *The sea power of the state*.

3

The Soviet Union and Jurisdictional Disputes in Northern Waters

Robin Churchill

INTRODUCTION

There are at the present time two major jurisdictional disputes involving the Soviet Union in Northern Waters. Firstly, the Soviet Union and Norway disagree over where the boundary should lie between their continental shelves and 200-mile economic zones in the Barents Sea. Secondly, there is a difference of opinion between Norway, on the one hand, and the Soviet Union and some other states on the other, as to whether the Spitsbergen Treaty, particularly its provisions giving all the parties to the Treaty the equal right to exploit Svalbard's economic resources, applies to the waters and sea-bed around Svalbard. The resolution of these disputes, which does not appear likely in the near future, will determine which states will be entitled to exploit the considerable resources, both living and non-living, of the areas in dispute and which states will be responsible for the management of the living resources of the areas concerned. The outcome of these disputes will also have considerable strategic and environmental implications.[1]

While the two disputes are obviously geographically and politically fairly closely related, they are nevertheless clearly two distinct questions, and in particular raise two different sets of legal issues. This chapter will therefore examine the two disputes separately. It will in each case look at the source of the dispute and the views of the main protagonists, and go on to indicate the significance of the possible solutions to each dispute for resources and security.

MARITIME BOUNDARY PROBLEMS IN THE BARENTS SEA[2]

Under the 1958 Continental Shelf Convention (to which both Norway and the Soviet Union are parties) the continental shelf is defined as the sea-bed out to a depth of 200 m (metres) or beyond if the depth of the superjacent waters admits of the exploitation of the resources of the sea-bed. Under the 1982 UN Convention on the Law of the Sea, the continental shelf is defined as the sea-bed out to 200 miles or the edge of the continental margin, whichever is the further. Although the 1982 Convention is not yet in force, the International Court of Justice has suggested that the definition of the continental shelf by the Convention has already passed into customary law.[3] Thus, under the 1958 Convention at least part of the bed of the Barents Sea is continental shelf — that part lying in less than 200 m of water, together with much beyond, however, that is exploitable. Under the UN Convention and customary law there is no doubt that the whole of the bed of the Barents Sea is, legally speaking, continental shelf. This means that since the emergence of the continental shelf as a concept in international law in the 1950s, there has been the need to establish a boundary between the continental shelves of Norway and the Soviet Union — initially, perhaps, for only part of the Barents, now undoubtedly for the whole Sea.

Norway has taken the view that the boundary should be the median line (i.e. a line equidistant from the nearest part of each state's mainland or insular territory). The Soviet Union has argued (at least in the earlier stages of the dispute) that the boundary should be a sector line, i.e. the line of longitude running from the terminus of the Norway–Soviet land frontier to the North Pole, modified so as to avoid passing through the Svalbard archipelago (see Figure 3.1). The sea-bed lying between the median line and the sector line is an enormous area of approximately 45,000 square nautical miles (nm^2) — an area greater than the Norwegian sector in the North Sea. Since 1974 several rounds of negotiations over the course of the boundary have taken place, apparently without a great deal of narrowing of each side's position, although the Norwegian government has made it clear that it would be prepared to modify its position on the median line in return for some concessions on the Soviet Union's sector claim. So far the Soviet Union does not appear to have indicated any willingness to make such concessions. The justification for each party's position will be examined in further detail below.

Since 1977 negotiations over a continental shelf boundary have become further complicated by the establishment by both Norway and

Figure 3.1: Disputed areas in the Barents Sea

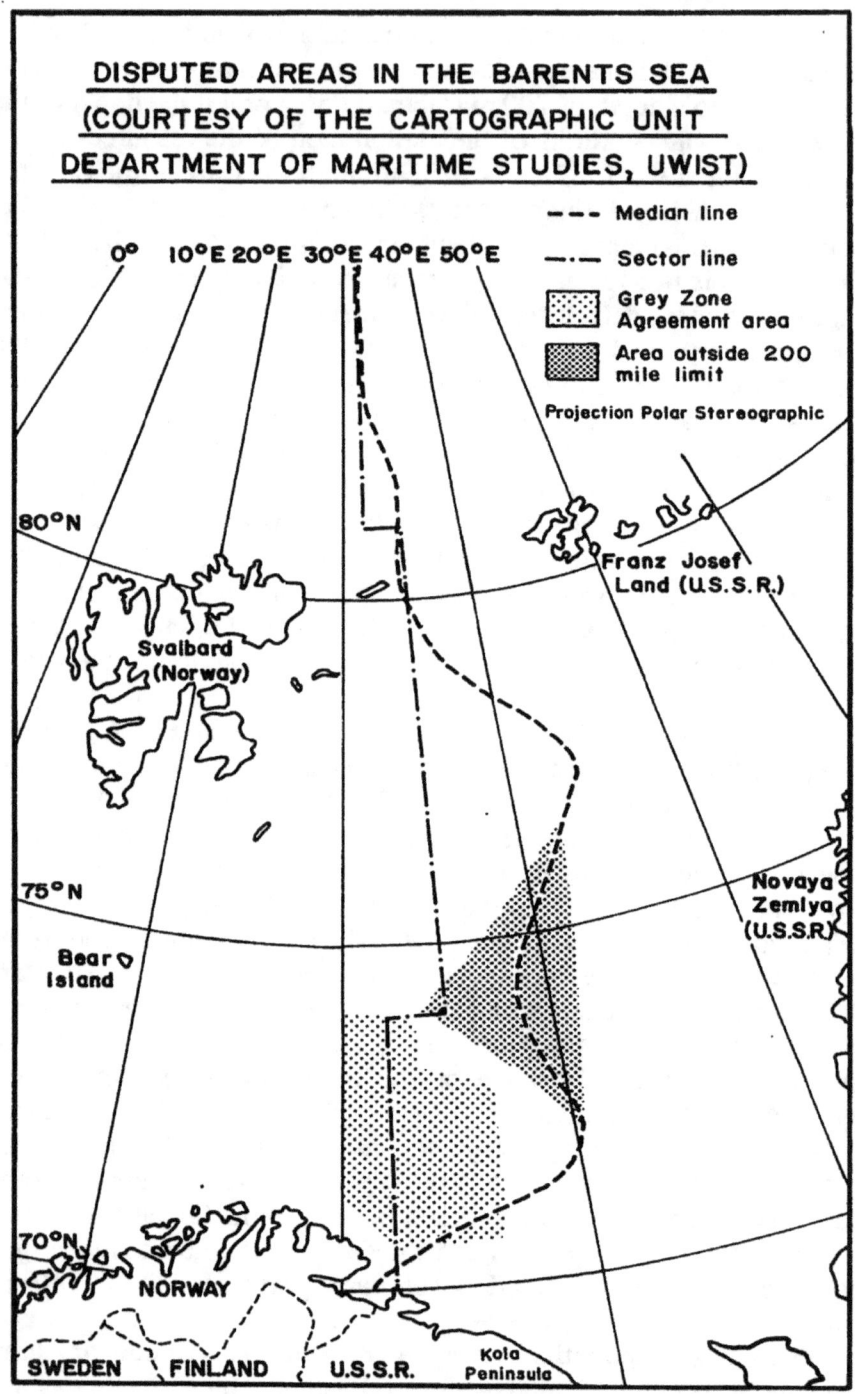

the Soviet Union of 200-mile economic zones.[4] This development meant that future negotiations would be concerned not just with a continental shelf boundary but also with an economic zone boundary. In the late 1970s there was no immediate urgency for a continental shelf boundary, as neither Norway nor the Soviet Union desired to begin early exploration for hydrocarbons in the disputed area. The same was not true for an economic zone boundary, however, as the southern Barents has been long and heavily fished by both Norwegian and Soviet fishermen. Neither Norway nor the Soviet Union wished speedily to establish an economic zone boundary (nor would such a boundary have been readily agreed), as this would largely have pre-empted the negotiations over the continental shelf boundary. Both sides were agreed, however, on the need quickly to come to some temporary practical arrangement for the exercise of each state's fisheries jurisdiction in the waters lying over the disputed area of continental shelf, i.e. the area between the median line in the east and the sector line in the west and within 200 miles of the mainland. This shared point of view led to the signature of the so-called 'Grey Zone Agreement' in January 1978.[5]

The agreement applies to an area which not only covers a large part of the disputed continental shelf area (the grey zone proper), but also to areas which are indisputably Norwegian or Soviet, i.e. areas to the west of the sector line and east of the median line respectively (see Figure 3.1) Within this area total allowable catches for the various fish stocks are to be decided by the Norwegian–Soviet Fisheries Commission established by an earlier fishery agreement of 1975. Total allowable catches are then to be allocated roughly equally between the two parties, with some quotas being allocated to third states after mutual consultation. The agreement also sets out various other regulatory measures (e.g. as to fishing gear and minimum fish sizes) which are to be observed by all vessels fishing in the area. Finally, the agreement provides that each party may exercise jurisdiction only in respect of its own fishing vessels and not in respect of vessels of the other party: jurisdiction over the fishing vessels of third states is to be exercised by whichever party has licensed such vessels.

The agreement is a temporary one, and was originally concluded for one year only. It has subsequently been extended for annual periods in every year following its signature, and this is likely to continue until agreement is reached on a continental shelf/economic zone boundary. In general the agreement has worked well, and it is some measure of its perceived usefulness and success that its annual renewals have continued uninterrupted, notwithstanding the fact that the non-socialist parties in Norway who were originally opposed to it replaced the Labour Party as the government for some time and the fact that a prominent member of the Norwegian delegation that negotiated the agreement, Arne Treholt, was subsequently convicted (in 1985) of being a Soviet spy.

Having examined the origins of the dispute over the continental shelf and economic zone boundaries (and the temporary solution to the absence of an economic zone boundary), we must now turn to consider what the solution to the dispute might be and, in particular, the arguments of the parties as to where the boundary should run in the light of the applicable international law.

At the outset it is desirable to recognise that the disputed boundary has three distinct segments, though whether the parties so view it is not known. The first segment runs from the termination of the existing maritime boundary (largely a territorial sea boundary)[6] at the mouth of Varangerfjord to a point 200 miles from the mainland of either Norway or the Soviet Union (or possibly a point 200 miles from both). In this segment there is a need for both a continental shelf and economic zone boundary between the adjacent mainland coasts of Norway and the Soviet Union. The second segment comprises the area in the middle of the Barents Sea which is more than 200 miles from any land. Here, only a continental shelf boundary is required. In this segment the relevant coasts of Norway (in part Svalbard, in part the mainland) and the Soviet Union (Novaya Zemlya) are essentially opposite one another. In the final segment, in the northern Barents, there is need once again for both a continental shelf and economic zone boundary. Here the relevant coasts — Novaya Zemlya and Franz Josef Land on the Soviet side, Svalbard on the Norwegian side — are again opposite. Each of the three segments must be considered separately, both because the applicable law differs and because different factors relevant to delimitation apply to each.

We will begin with the southern segment, which in practice is the area where the need for a boundary is greatest. Before 1977, negotiations in this area were concerned solely with a continental shelf boundary. For this the applicable law was the Continental Shelf Convention (since both states are parties to it). Article 6 of the Convention provides that in the absence of agreement, the boundary line is to be the median line unless another line is justified by special circumstances. Since 1977 it appears (though like much else in the negotiations, little has been said publicly) that the negotiations have become concerned with seeking to establish a single boundary for both the economic zone and continental shelf.[7] If this is so, then according to the International Court of Justice in the *Gulf of Maine* case,[8] the applicable law is no longer Article 6 of the Continental Shelf Convention but customary international law. According to the Court, customary law lays down the basic principle that 'delimitation is to be effected by the application of equitable criteria and by the use of practical methods capable of ensuring, with regard to the geographic configuration of the area and other relevant circumstances, an equitable result',[9] but does not prescribe any particular 'criteria' or 'practical methods' for effecting a delimitation.

Norway takes the view that the most appropriate practical method for effecting a delimitation is the median line (or, in terms of the Continental Shelf Convention, that there are no special circumstances that justify a departure from the median line). The Soviet Union takes the view that there are factors which indicate that a median line is not the appropriate method of delimitation (or, in terms of the Continental Shelf Convention, that there are special circumstances). The factors that the Soviet Union apparently has in mind are its greater size, the greater population of the Kola Peninsula as compared with northern Norway, and its security interests.[10] As regards the first of these, this is not a factor that an international court would be likely to consider relevant: in the *Continental Shelf (Libya/Malta)* case, for example, the International Court of Justice rejected Libya's argument that a state with a greater land mass should have a greater share of continental shelf.[11] However, if by its greater size what the Soviet Union is really getting at is the fact that its coastline in the southern Barents is considerably longer than that of Norway, each coast being measured according to its general direction and not in all its sinuosities (as international courts prescribe), then it is on much firmer legal ground. International courts have stressed several times that there should be a reasonable (though not exact) degree of proportionality between the maritime areas delimited and the length of the parties' respective coastlines.[12] Thus, it seems that the Soviet Union would be justified in raising the greater length of its coastline as a relevant factor to be taken into account in the delimitation. The other factors alleged by the Soviet Union are its greater population and security. The trend of the International Court's decisions is to exclude as irrelevant to delimitation all factors except the geographical, particularly where, as is likely to be the case here, the boundary line relates to both the continental shelf and economic zone. On the other hand, in both the *Guinea/Guinea Bissau* arbitration[13] and the *Continental Shelf (Libya/Malta)* case[14] security factors were not dismissed out of hand as irrelevant, though in neither case was it made very clear in what circumstances they might be relevant: what the Court in each case seems to have had in mind is that a state should have control of the maritime territory immediately off its coast. In that sense, security factors would seem to reinforce the non-encroachment principle (discussed below).

There is a further factor which has not apparently been raised by the Soviet Union but which might be relevant. The Norwegian coast projects a little further seawards than the Soviet coast along what is basically a straight coast between the White Sea and the North Cape. The effect of this is to push the median line a a little further to the east, possibly to such an extent that it might be regarded as encroaching on maritime areas that more naturally belong to the Soviet Union. The principle of non-encroachment is one that has been widely recognised by international courts.[15]

It was pointed out earlier that the Soviet Union has apparently argued that the continental shelf boundary should be a sector line. The actual line of longitude to which the Soviet Union refers was that used in a Soviet decree of 1926 to define the western limit of its claim to *land* (not maritime) territory in the Arctic. It is not clear whether the Soviet Union regards the sector line as the line which results from the existence of the special factors discussed above or whether the sector line itself is regarded as a relevant factor. The former seems unlikely, as it would be highly artificial to consider that the relevant factors should give rise to such a precise and, from the Soviet point of view, well-established line. The latter possibility also seems unconvincing, for a sector line is not of the character of the factors which international courts have hitherto recognised as relevant. It is also possible that the Soviet Union considers that use of the sector line has a justification quite separate from the ordinary law of maritime boundary delimitation, namely that a sector line is the appropriate way of drawing a maritime boundary in polar areas. If this is its contention, the Soviet Union would have to prove that the use of sector lines as a method of drawing maritime boundary lines in polar areas is a rule of customary international law. This would seem impossible to prove, because there is both insufficient practice and inadequate *opinio iuris* (state practice and *opinio iuris* being the two constituent elements of customary international law). It is true that sector lines have been used as boundaries in the Antarctic, but this use is limited to land territory, and seems to be motivated by reasons of convenience, not because the states concerned felt themselves to be under any legal obligation to use such lines.[16] In the Arctic Norway and the USA have protested against the use of sector lines, and the Soviet Union is the only state which has used such lines, although its practice as regards maritime areas has by no means always followed the sector theory.[17] In any case an international court would probably regard the sector line as being an inequitable solution in the Barents because it would encroach on maritime areas which more properly belong to Norway.

In relation to the first segment of the maritime boundary in the Barents, the most important factors would seem to be that Norway and the Soviet Union are adjacent states along what is essentially a straight coastline and that the Soviet coastline in the southern Barents is considerably longer than the Norwegian coastline. In this situation neither the median line nor the sector line appears to be the appropriate method or constitutes an equitable solution. What might be regarded as an equitable solution would be to draw the boundary line perpendicular to the general direction of the coast. This was commended by the International Court in the *Gulf of Maine* case as a suitable method of delimitation on coasts which are more or less straight.[18] Furthermore, a perpendicular line would respect the principles of non-encroachment and proportionality.

Having looked at the first segment, we must now turn to look at the second, in the middle of the Barents. Here, it will be recalled, only a continental shelf boundary is required. Thus, the applicable law will be Article 6 of the Continental Shelf Convention. In other words, the boundary will be the median line unless another line is justified by special circumstances. Of the factors raised by the Soviet Union and discussed above, the sector line is clearly not a special circumstance. Nor is the principle of non-encroachment relevant where, as here, the respective coasts are opposite rather than adjacent to one another. While it is difficult to calculate the length of the relevant Norwegian coastline (part of Svalbard, and possibly part of the mainland), it does not seem excessively disproportionate when compared with the relevant Soviet coastline (part of Novaya Zemlya). What possibly might amount to a special circumstance, therefore requiring some adjustment to the median line, is the fact that two of the islands in the Svalbard archipelago — Hope Island and Bear Island (and the latter may not be relevant) — lie a long way from the main part of the archipelago.

In the third and final segment there is, as with the first segment, the need for both a continental shelf and economic zone boundary. Thus, the applicable law will be customary law, and not the Continental Shelf Convention. Unlike in the first segment, the coasts of the parties here are opposite, not adjacent to one another. This means that the principle of non-encroachment would not seem to be relevant. Furthermore, again unlike in the first segment, there does not seem to be an excessive disproportion between the lengths of the parties' coastlines. In general international courts have taken the view that equidistance is an appropriate method for delimitation (at least as a starting point) in the case of opposite coasts.[19] Thus, it would seem that in the present case equidistance would in principle be the appropriate method of delimitation, though some adjustment might be thought desirable to discount (wholly or in part) some of the outlying islands of the Svalbard and Franz Josef Land archipelagos.

The means by which a solution to the boundary dispute in the Barents may be reached are either continued negotiations between the parties or recourse to an international judicial or arbitral tribunal. The latter is most unlikely, given the Soviet Union's long-standing opposition to the use of such tribunals for solving disputes in which it is involved. Thus, it seems that continued bilateral negotiations will be the medium used to try to reach a solution. It should be stressed that while some suggestions have been made above as to the possible course of the boundary indicated by international law, it is open to the parties, whether the applicable law is the Geneva Convention or custom, to agree on any boundary line they wish. The discussion above of the relevant law, particularly the case law of international courts, is not, however, without interest or relevance. It is likely that various aspects of this law are quoted by the parties in the negotiations in order to support a particular line of argument.

The speed at which negotiations over a maritime boundary in the Barents are continued and the likelihood of their reaching a successful conclusion depend to a considerable extent on how strongly the parties feel the need to explore new areas for sea-bed hydrocarbons. (As long as the 'grey zone' agreement remains in force, fisheries interests will not press either party to seek an early solution to the boundary problem.) At the moment both Norway and the Soviet Union appear to be reasonably well endowed with hydrocarbon resources elsewhere, and nor is the Barents yet apparently of any great commercial interest to the oil companies. However, it may be significant that since 1983 the Soviet Union has conducted exploratory drilling just to the east of the disputed area. It is also noteworthy that at their December 1984 round of negotiations the parties apparently agreed that they would not conduct exploratory activities within the disputed area so long as the boundary is not agreed.[20] Experience elsewhere in the world (particularly in the North Sea) suggests that it will be easier for Norway and the Soviet Union to agree on a continental shelf boundary if the hydrocarbon resources of the disputed area are not identified beforehand. While the respective needs of the parties for hydrocarbons constitute a major factor in determining the course of negotiations, they are not the only factor. For the Soviet Union, security considerations are also very important. Its security interests would, in the absence of any pressing need for resources, lead the Soviet Union to delay negotiations. So long as there is no agreed boundary, the Norwegians will not drill in the disputed area or in all probability very close to it. Wherever a boundary is located, it will almost certainly lead to Norwegian offshore activity pushing eastwards, something which is inimical to Soviet security interests: the further east Norwegian offshore activity takes place, the greater the possibility of Norwegian drilling rigs obstructing or monitoring Soviet naval vessels on their way into or out of the ports of the Kola Peninsula.[21] A final factor in determining the speed of negotiations is the fact that the Soviet Union may perceive that Norway appears to be keener on a boundary agreement than it is: it can therefore afford to wait, hoping that this will lead to concessions by Norway.[22]

If the negotiations achieve a positive result, it will probably be in the form of agreement on a single boundary line. A theoretical alternative is to abandon the search for a boundary and to provide instead for a regime of joint use of the disputed area (as, for example, Sudan and Saudi Arabia have done in the Red Sea). It seems unlikely that such a solution would appeal to either of the parties (although the Soviet Union has recently indicated some interest in co-operative arrangements). If the solution is, as is most likely, a single boundary line, it may possibly be more readily achieved if the boundary agreement were to provide for the non-use of offshore installations for espionage purposes (although this may raise difficult questions of verification and inspection) and for non-obstruction by installations of either party's vessels. For the sake of fisheries interests and the environment generally, it is to be hoped that a boundary agreement will continue the present arrangements for the management of joint fish stocks and contain meaningful provisions on pollution control.

THE LEGAL REGIME OF SVALBARD'S MARITIME ZONES

The jurisdictional dispute involving Svalbard is of a different nature from the Barents Sea dispute just discussed. Whereas the latter is concerned with the boundary between the maritime zones of Norway and the Soviet Union, the Svalbard dispute is concerned with the question of whether the legal regime laid down by the 1920 Treaty of

Spitsbergen[23] — particularly its provisions giving all 40 or so states party to the Treaty an equal right to engage in certain economic activities (including fishing and mining) on Svalbard — extends beyond the land territory and territorial sea of Svalbard (both of which are expressly covered by the Treaty) to the maritime zones beyond, namely the continental shelf and economic zone.[24] Norway takes the view that the Treaty does not apply beyond the territorial sea. The Soviet Union and one or two other states who are parties to the Treaty take the view that the Treaty does apply. The majority of states parties, including the USA, United Kingdom and most West European states, have not publicly expressed a view either way, but have reserved their position.

If the Treaty does not apply beyond the territorial sea, very different consequences follow than if the Treaty does apply. If the Treaty does not apply, then Norway will have the exclusive right to exploit the economic resources of Svalbard's continental shelf and 200-mile zone, subject only to the general international law obligations applying to coastal states in the exploitation of their offshore resources. On the other hand, if the Treaty does apply, all 40 or so states parties to the Treaty will have an equal right to exploit the rich fishery resources in the waters around Svalbard (which, it is important to remember, include not only the main Spitsbergen archipelago, but also Bear Island, lying about half-way between the North Cape and the main archipelago). All states parties will also have an equal right to explore Svalbard's continental shelf for hydrocarbons, and if any oil or gas is found (the possibilities for which are thought to be quite promising), the right to exploit it. Such a right of exploitation would benefit from an extremely liberal taxation regime, because under Article 8 of the Spitsbergen Treaty, Norway may only tax economic activities on Svalbard at a level which does not produce more revenue than is required for the needs of Svalbard itself. Any significant offshore petroleum activity on Svalbard's continental shelf, especially by Western oil companies, is likely to be regarded by the Soviet Union as prejudicial to its security, since such activity would increase the possibilities of monitoring the movements of Soviet naval vessels in the Barents, Norwegian and Arctic Seas. Any significant offshore petroleum activity also has environmental implications, because the more such activity, the greater the risk of pollution in waters which are not only rich in marine life but also ecologically delicate (because of the length of time which it takes for bacteria to break down oil in cold waters).

Having seen that there is a difference of view between Norway and the Soviet Union over whether the Spitsbergen Treaty applies beyond the territorial sea of Svalbard, we must now turn to examine the legal arguments relevant to each party's position.[25] To support its view that the Treaty does not apply, the Norwegian government puts forward two main arguments. Firstly, it says that there is a continuous continental shelf extending northwards from North Norway on which the Svalbard archipelago sits. Norway's rights to this continental shelf derive from its sovereignty over North Norway, not Svalbard: in other words, this is North Norway's continental shelf, not Svalbard's. Since this is not Svalbard's continental shelf, the Treaty does not therefore apply. The weakness of this argument is that it ignores the rule of international law that every island (except possibly uninhabitable rocks) has its own continental shelf. From this rule it follows that Svalbard has its own continental shelf. The question, then, is whether the Treaty applies to this continental shelf, and this is a question that can only be answered with other arguments. The Norwegian argument mentioned here in effect assumes that Svalbard's continental shelf is under the same legal regime as mainland Norway's continental shelf, i.e. it assumes the very thing to be proved. A further weakness with this first Norwegian argument is that if it were correct, it would effectively deny Svalbard's capacity to generate a continental shelf. It is then difficult to see how Svalbard could generate a fishery protection zone or serve as a basepoint for the continental shelf boundary with the Soviet Union — both of which Norway has claimed.

The second Norwegian argument is that the Spitsbergen Treaty should be given a literal and restrictive interpretation. The Treaty does not mention the continental shelf or economic zone, therefore it cannot apply to them. Furthermore, to say that the Treaty applied to Svalbard's continental shelf and economic zone would be to impose on Norway — which has sovereignty over Svalbard — an obligation which is not expressly mentioned in the Treaty: it is a well-known rule of treaty interpretation that restrictions on sovereignty are not to be presumed. This second Norwegian argument is stronger than the first, but is not wholly convincing. A literal reading of the Treaty is not free of difficulties. The geographical extent of the equal right of fishing given by the Treaty is 'the territories specified in Article 1' (namely, various islands listed by name, and all other islands within a defined area) and 'their territorial waters' (Article 2), while the geographical extent of the equal right of mining is 'the waters, fjords and ports of the territories specified in Article 1' and 'on land and in the territorial waters' (Article 3). Some of these terms ('territories', 'waters') are not very precise, while the term 'territorial waters' was unclear in international law at the time when the Spitsbergen Treaty was drafted. Since a literal reading of the Treaty is not free of difficulties, it is also necessary to have regard to the object and purpose of the Treaty (to maintain other states' previous liberal access to Svalbard no less than to recognise Norwegian sovereignty) and the circumstances of the Treaty's conclusion (the uncontested *terra nullius* status of Svalbard prior to the Treaty's conclusion and Norway's initial reluctance to acquire Svalbard). These factors cast further doubt on a literal interpretation, and they also argue against a restrictive interpretation of the Treaty, a method of interpretation which in any case is not generally favoured in international law today.

In sum, therefore, the Norwegian arguments for the Spitsbergen Treaty not applying beyond the territorial sea are not wholly convincing and do not seem conclusive of the issue. What are the arguments to support the view that the Treaty does apply beyond the territorial sea? The Soviet Union and other states taking this view have not made their arguments public, but it is likely that the following arguments are those that they have in mind. Firstly, Norwegian sovereignty over Svalbard is restricted and subject to limitations. Since the Norwegian right to maritime zones around Svalbard derives from its sovereignty over Svalbard (and international courts and tribunals have frequently stressed that a coastal state's right to maritime zones derives from its sovereignty over the adjacent land), then it should follow that Norway's rights in the maritime zones around Svalbard are also subject to limitations. Secondly, if the Treaty did not apply beyond Svalbard's territorial sea, anomalies would be created in that parties to the Treaty would have an equal right of economic exploitation on Svalbard's land territory and territorial sea, but no such right in Svalbard's zones seawards of the territorial sea. Thirdly, the Treaty should be extended by analogy. When it was concluded, the Treaty applied to the one maritime zone then known, the territorial sea. By analogy it should be extended to the maritime zones which have subsequently evolved in international law. This, it is reasonable to assume, would have been the intentions of the drafters of the Treaty if they had known about these newer maritime zones. A final argument would be that the Treaty should be interpreted so as to give emphasis to its object and the circumstances of its conclusion mentioned above.

Taken together, these arguments do seem stronger than the Norwegian arguments but are not perhaps of such a nature as to be overwhelmingly conclusive. In any case, this dispute is unlikely to be resolved by purely legal arguments. That would only be the case if the dispute were referred to an international judicial tribunal, something which is extremely unlikely given the Soviet Union's already-mentioned dislike of using such bodies for settling disputes. (Of course, it is not inconceivable that a Western state might decide to move from its present position of public fence-sitting to a position of opposition to the Norwegian view and refer the dispute to the International Court, but given the security issues in the Svalbard area, this again seems unlikely.) Thus, it seems that if this dispute is to be resolved, it will be by diplomatic means, by direct negotiations between Norway and the other parties to the Spitsbergen Treaty. This could be done either by holding a conference of the parties to the Spitsbergen Treaty or by some less formal means. The latter seems preferable, for a formal conference might simply lead to an entrenching of existing views and also risks opening up questions relating to Svalbard that are well settled.

If the dispute is to be resolved by diplomatic means, what should the solution be? In this writer's view, the better solution would be a recognition that the Treaty should not apply beyond Svalbard's territorial sea. This view is based on strategic and environmental considerations. If Svalbard's continental shelf were not subject to the Treaty, it is likely, given the cautious approach of Norway to oil exploration north of the Arctic Circle, Norwegian economic interests and a Norwegian desire to minimise disturbances to the *status quo* in a strategically sensitive area, that oil exploration and exploitation of Svalbard's continental shelf would proceed at a slower, possibly much slower, rate than if the Treaty did apply. This would not only reduce the risk of pollution from sea-bed activities and conflicts between fishing and the offshore oil industry, but would also reduce the potential for surveillance of Soviet naval vessels from sea-bed installations. Furthermore, even though non-Norwegian oil companies might be licensed to explore Svalbard's continental shelf, Norway would be in a stronger position to offer guarantees for the non-use of the installations of such companies for espionage and to reduce their potential for obstructing the passage of Soviet vessels than if the Treaty did apply.

There is an alternative to the 'either/or' choice of the Treaty applying beyond Svalbard's territorial sea, and that is the elaboration of a special legal regime for Svalbard's maritime zones beyond the territorial sea. Lack of space precludes discussion of how such a regime might be elaborated and what its content might be, but the present fishery protection zone does give some indication of how such a regime might be elaborated and what one of its constituent elements might be.

Before leaving the question of Svalbard, it is of interest to speculate as to why the Soviet Union has publicly taken the view that the Spitsbergen Treaty applies beyond Svalbard's territorial sea. At first sight, this position seems contrary to its own interests because the Soviet Union can hardly relish Western oil companies exploring the sea-bed around Svalbard. There are a number of possible explanations for the Soviet stance. It may be that the Soviet Union hopes that by publicly opposing Norway over Svalbard, it can later relax its opposition in return for concessions by Norway on the Barents Sea maritime boundary question.[26] Secondly, the Soviet Union may be motivated by the fact that as long as there is a dispute over the application of the Spitsbergen Treaty to Svalbard's continental shelf, there will be no exploration or exploitation of that shelf — something which is in the Soviet interest for reasons mentioned earlier. Thirdly, it may be that the Soviet Union has taken its present position as a safeguard for the future, in case its need for oil becomes so great that it wishes to explore the Svalbard shelf itself. Whatever the explanation, it is worth remembering that the Soviet Union's present stance is consistent with the policy it has practised since the 1940s of challenging Norwegian sovereignty over Svalbard whenever the opportunity presents itself.

CONCLUSIONS

The disputes over a maritime boundary in the Barents Sea and over the legal status of Svalbard's maritime zones beyond the territorial sea have been in existence for well over a decade, and show no signs of an early settlement. As long as these jurisdictional disputes continue, there will be less than optimum management and exploitation of the resources of the areas concerned. There may also from time to time be an unsettling effect on the general security of the region, particularly if one party engages in acts designed to assert its views as to how the disputes should be solved, as, for example, when the Soviet Union engaged in missile tests in the disputed continental shelf area in the southern Barents in 1976. For most of their existence, however, the disputes have been kept as low-key affairs, certainly as a matter of deliberate policy by Norway and probably also by the Soviet Union. A solution to one or both disputes would only be an improvement on the present situation if it provided equal or better management of resources, greater stability to the security of the region, and effective environmental controls on the increased offshore petroleum activity that would be likely to result. On the other hand, if there is no solution to the disputes, it will probably not be possible to maintain their present low-key nature indefinitely because of the pressure that will almost certainly eventually build up to exploit the sea-bed resources of the areas in dispute.

NOTES

1. For details of the resources and strategic significance of the areas concerned, see Chapters 5 to 8 inclusive.
2. This section of the chapter is based on and largely taken from R.R. Churchill, 'Maritime boundary problems in the Barents Sea' in G.H. Blake (ed.), *Maritime boundaries and ocean resources* (Croom Helm, Beckenham, 1987).
3. *Case concerning the Continental Shelf (Libya/Malta)*[1985] International Court of Justice (ICJ) Rep. 13 at 33. In its domestic legislation the USSR uses the Continental Shelf Convention definition (see Decree of 6 February 1968: text in UN Legislative Series B/15, p. 441). Norwegian law originally defined the continental shelf as being the sea-bed out to the depth of exploitability (see Royal Decree of 31 May 1963:text in ibid., p. 393), but the definition has now been changed to that of the UN Convention (see the Petroleum Law of 22 March 1985).
4. The Norwegian economic zone was established by Law No. 91 of 17 December 1976, text in UN Legislative Series B/19, p. 241. The Soviet Union originally established a 200-mile fishing zone (see Decree of 10 December 1976, text in ibid., p. 253), but replaced this in 1984 with a 200-mile economic zone (see Decree of 28 February 1984, text in *Law of the Sea Bulletin*, no.4 (1984), p. 32).
5. The official title of the agreement is: Agreement on an Interim Practical Arrangement for Fishing in an Adjoining Area in the Barents Sea. An English translation of this agreement does not appear to have been published; the Norwegian text can be found in [1978] *Overenskomster med fremmede Stater*, 436.
6. Established by the agreement concerning the Sea Frontier between Norway and the USSR in the Varangerfjord, 1957. Text in UN Treaty Series vol. 312, p. 289.
7. This is certainly the implication given by the then Norwegian Prime Minister in a speech made in February 1978: see *UD—informasjon*, no. 7 (1978).
8. [1984] ICJ Rep. 246.
9. Ibid., pp. 299–300.
10. As mentioned above, few details of the negotiations have been made public. The factors mentioned here have been indicated by the Norwegian government as being the ones the USSR has in mind: see *UD—informasjon*, no. 30 (1977).
11. *Case concerning the Continental Shelf (Libya/Malta)*[1985] ICJ Rep. 13 at 40–1. Note, too, that in the *Guinea/Guinea Bissau Maritime Boundary* Arbitration (text reproduced in (1985) 89 *Revue Générale de Droit*

International Public, 484), the Court said that the respective size of the parties' land territory was irrelevant (paras 118—19).

12. See *North Sea Continental Shelf* cases [1969] ICJ Rep. 3 at 52; *Continental Shelf (Tunisia/Libya)* case [1982] ICJ Rep. 18 at 75; *Gulf of Maine* case [1984] ICJ Rep. 246 at 322–3, 334–7; *Guinea/Guinea Bissau* Arbitration, paras 118–20; *Continental Shelf (Libya/Malta)* case [1985] ICJ Rep. 13 at 43–6, 49–50.

13. Para. 124.

14. [1985] ICJ Rep. 13 at 42.

15. See *North Sea Continental Shelf* cases [1969] ICJ Rep. 3 at 31–2; *Continental Shelf (Tunisia/Libya)* case [1982] ICJ Rep. 18 at 61–2; *Gulf of Maine* case [1984] ICJ Rep. 246 at 298–9, 313, 328; *Guinea/Guinea Bissau* Arbitration, paras. 103–7.

16. See D.J. Harris, *Cases and materials on international law*, (Sweet and Maxwell, London, 1983), p. 181.

17. Ibid., p. 183; W. Østreng, 'Delimitation arrangements in Arctic seas', *Marine Policy*, no. 10 (1986), pp. 144–8. Canadian practice concerning sector lines in the Arctic is ambivalent. See D. Pharand, *The Law of the Sea of the Arctic* (University of Ottawa Press, Ottawa, 1973), pp. 134–41, and D. Pharand, 'The implications of changes in the Law of the Sea for the "North American" Arctic Ocean' in J. K. Gamble (ed.), *Law of the Sea: neglected issues* (Law of the Sea Institute, Hawaii, 1979), p. 184.

18. [1984] ICJ Rep. 246 at 320. A similar view appears to have been taken by the Court in the *Tunisia/Libya* case.

19. See the *North Sea Continental Shelf* case [1969] ICJ Rep. 3 at 36–7; *Anglo-French Continental Shelf* arbitration, Cmnd 7438, para. 239; *Continental Shelf (Tunisia/Libya)* case [1982] ICJ Rep. 18 at 88; *Gulf of Maine* case [1984] ICJ Rep. 246 at 334; *Continental Shelf (Libya/Malta)* case [1985] ICJ Rep. 13 at 47.

20. *The Times*, 4 December 1984. Whether international law requires a state to abstain from drilling in a disputed area of continental shelf is not a question to which a straightforward answer can be given, and lack of space precludes any discussion of this point.

21. On the other hand, prolonged continuation of the boundary dispute may lead to increased tension in the area, which is not in the Soviet Union's security interests: see Østreng, 'Delimitation arrangements', p. 149.

22. This point was made by Traavik and Østreng as long ago as 1977, but still appears to be valid, although it may be that the Soviet Union's petroleum needs will not make it valid for so very much longer. See K. Traavik and W. Østreng, 'Security and ocean law: Norway and the Soviet Union in the Barents Sea', *Ocean Development and International Law*, no. 4 (1977), p.358. Norway appears to be just as keen on an agreement as when Traavik and Østreng wrote, since it takes the view that an agreement would lead to greater stability in the region—something which has been the policy of successive Norwegian governments.

23. Text of the Treaty in League of Nations Treaty Series, vol. 2, p. 8. The Treaty gave Norway sovereignty over Svalbard, which had previously been regarded as a no-man's-land, subject to certain conditions, including the obligation to permit other parties to the Treaty to engage in certain economic activities on an equal basis.

24. At the present time Svalbard has a continental shelf and a 200-mile fishery protection zone, established in 1977. A fishery protection zone (in which Norwegian fishermen enjoy no advantages compared with the fishermen of other states parties to the Spitsbergen Treaty) was chosen in preference to a normal economic zone (which would have given Norwegian fishermen advantages) in order not to prejudice the question of whether the Treaty applies to the waters embraced by a 200-mile zone.

25. For a fuller discussion of the legal arguments than that which follows, see R.R. Churchill, 'The maritime zones of Spitsbergen' in W.E. Butler (ed.), *The Law of the Sea and international shipping: Anglo-Soviet post UNCLOS perspectives* (Oceana Publications, New York, 1985), pp. 189–234.

26. For a fuller discussion of a package deal approach to resolving the Svalbard and Barents delimitation problems, see Østreng, 'Delimitation arrangements', pp. 153–4.

4

The Maritime Policies and Practices of The Soviet Union after UNCLOS III, With Special Reference to Northern Waters

Uwe Jenisch

Among the five circumpolar states, the Soviet Union clearly has a dominating position in terms of geography, military strength, sea traffic and probably also in the field of potential resources.

OLD AND NEW SOVIET USES IN THE ARCTIC

Maritime navigation along the difficult Arctic routes is mainly based on the use of powerful ice-breakers, the most important of these being nuclear vessels. The Soviet ice-breaker *Arktika* was the first surface ship in August 1977 to reach the North Pole. In Finland the Soviet Union is building a new generation of Arctic multi-purpose ships with ice-breaking capabilities to deliver supplies to the Soviet coasts and islands in the Arctic and to open up the Northeast Passage[1] between the Barents Sea and the Bering Strait on a regular basis. This route, covering 2,800 km, is open to navigation for between two and four months a year.

Soviet sources[2] predict that by the beginning of the next century, Arctic sea routes will transport more cargo than the new BAM (Baikal–Amur–Magistrale) Railway. When in October 1983 more than 80 Soviet ships were entrapped by ice in East Siberian waters, the world became aware of the intensity of Arctic shipping — and of its risks. Year round navigation is hoped to be achieved by the early 1990s. Maritime access and transportation are the key problems for the future development of Arctic resources and the Soviet Union and Canada have clearly taken a leading role here.

Systematic scientific research conducted by Arctic and other states[3] is being intensified to provide basic marine data. Meteorological observation and pollution control is becoming increasingly more important. Prevailing winds and currents carry pollutants northwards from the Eurasian and American continents. Regarding climatic conditions of the Northern Hemisphere, the Arctic is the glass roof of the greenhouse in which we live.

A typical freedom of the seas which has been exercised by US and Soviet scientists is floating research stations on ice shelves which drift with the Arctic currents regardless of any sector lines or limits of national jurisdiction. Once the Law of the Sea (LOS) Convention comes into force most of these projects will be regulated by the coastal states,[4] and one may fear that the need for permission by one or more Arctic states will prove to be a serious obstacle to future scientific research.

The main hydrocarbon deposits have been located on the Alaskan and Canadian shelves, on the Norwegian shelf, and especially in the Barents and Kara Seas.[5] The deposits often coincide with terrestrial deposits on Alaskan, Canadian, Norwegian (Spitsbergen) and Soviet territory.

Regarding the Soviet Union, political priorities could open up production some time in the near future, as the demand for oil for domestic consumption and for hard currency needs is still growing. Due to relatively favourable ice conditions the continental shelves of northern Norway and of the neighbouring Soviet Union, i.e. the Barents and Kara Seas, appear to be the natural starting-point for oil development.[6] Early in 1984, at the request of Soviet partners, a Norwegian consulting firm elaborated a development plan for the exploration of the Soviet part of the Barents Sea.

European geographical conceptions tend to underestimate the growing strategic importance of the Arctic. The two superpowers are both riparian states of this 'Arctic Mediterranean' which serves as a buffer zone and as a passageway. The polar route is the shortest communication line (3,500 km) between both states and it is easily accessible for submarines, aircraft and missiles. Shallow waters, uninhabited islands, and the icecover allow the emplacement of 'scientific' and military installations[7] for various purposes, for example, for detection and tracking of naval forces and of missiles.[8]

The industrial and population centres of both superpowers lie on a north–south course across the North Pole, thus providing 'missile avenues' for intercontinental ballistic weapons. Apart from this, the Soviet Union has increased the military importance of the area by concentrating army, airforce and navy forces in ice-free bases on the Kola Peninsula. Murmansk and Severomorsk are the home ports for about 40 per cent of the Soviet navy, including about 66 per cent of the maritime nuclear strike force. The Soviet Northern Fleet is one of the largest of all Soviet fleets, consisting of 39 ballistic-missile submarines, 117 conventional submarines, one aircraft carrier, 73 principal surface combatants and 45 minor surface combatants.[9] Submarine-based nuclear missiles can hit the whole Northern Hemisphere including North America, Western Europe and China, while the submarines stay in the Barents Sea or its neighbouring waters, thereby using the Sea as a 'stationing area' for missiles. Since 1976 the Soviet Union has used the Sea for the testing of missiles while negotiating with Norway over maritime border lines;[10] and again in July 1985, during the Soviet naval exercises, a 'warning area' (from 18–24 July for 4 hours/day) was proclaimed in the waters north-east of Jan Mayen. To sum up, these waters are the focus of East–West military confrontation. The Svalbard Passage (Spitsbergen Passage) between the north Norwegian coast and the island of Spitsbergen represents the Soviet Union's most important outlet and access to the oceans. The Soviet Union considers this an extremely strategic area and fears that because of its massive military build-up, it is vulnerable here. Norway and NATO have no equivalent forces in the area. Nevertheless all vessels going to and from Soviet Arctic ports have to pass through waters under Norwegian jurisdiction and can be closely monitored by NATO.

THE NEW MARITIME ZONES

Having the unique characteristic of being 80–90 per cent permanently frozen might lead to a certain confusion as to whether the Arctic Ocean is 'land' in the sense of *terra firma* and/or 'sea' in terms of international law. While academic Soviet sources[11] tend to profit from the terrainisation aspect and use it as a pretext for claims of exclusiveness, it should be maintained that the Arctic clearly is regulated by the international Law of the Sea.[12] Throughout the Law of the Sea Conference there were no objections against the full application of the new law to Arctic waters. The conference avoided any tendencies to establish a 'special' or 'regional' regime for polar seas, the only exception being Article 234, which dealt with 'ice-covered areas' in the context of the anti-pollution provisions of the convention. This article introduces special environmental rights for the prevention, reduction and control of marine pollution from vessels in ice-covered areas within the limits of the Exclusive Economic Zone (EEZ), provided that any such regulation must be 'non-discriminatory' and have 'due regard to navigation'. Except for the narrow circumstances of this article, freedom of navigation and all other freedoms are applicable.

Moreover, the same pattern of conflict of interest between neighbouring and opposite states over rights to resources and rights to control maritime access, which has been so characteristic for the whole LOS Conference, can also be seen in relation to the polar seas.[13]

Fortunately, there are practically no unsettled disputes[14] concerning sovereignty over continental or insular land territory since the Norwegian sovereignty over Svalbard was agreed upon in 1920 and the Permanent Court of International Justice settled the East Greenland conflict in 1933 in favour of Denmark. Thus, the lateral delimitation between neighbouring states, as well as the seaward limit of the continental shelf and the EEZ, are among the main aspects to be discussed in this chapter. The LOS Convention offers little help. After lengthy debates the delimitation regulations in Articles 83 and 74 for continental shelves and EEZs give priority to 'agreement on the basis of international law, as referred to in Article 38 of the Statute of the International Court of Justice, in order to achieve an equitable solution'. Instead of offering certain criteria for delimitation, everything is practically left to bilateral negotiations. Moreover, according to Article 298, disputes over delimitation questions can be exempted from the compulsory dispute settlement system — and in fact have been exempted by the Soviet Union upon signature.[15]

In view of the large number of delimitation problems in the Arctic, this ocean has already become a playground for various methods of delimitation such as the median line principle, the sector theory, special circumstances, equitable principle, and joint development zones.

NORWEGIAN-SOVIET BORDER PROBLEMS

The large continental shelf of the Barents Sea has been the object of protracted delimitation talks between Norway and the Soviet Union since 1967 and there is little prospect of a solution in the near future. Negotiations have been intermittent since 1981.

The contested area[16] comprises particularly rich fishing banks, is a potential source of oil and gas, and is situated in the strategically important approaches to the Kola Peninsula. The average depth in the Barents Sea, including the waters around Svalbard, is only 229 m with a maximum of about 500 m in the Barents Trough south of Bear Island. The geographical details and the negotiating positions have been well discussed in detail in the literature.[17] Norway maintains that the median line, as laid down in Article 6 of the 1958 Geneva Shelf Convention, should apply [18] while the Soviet Union insists upon 'special circumstances' indicating sector-line division along 32° 04' 35" E. The difficulties arise from the simple fact that Article 6 of the Shelf Convention provides for the median or equidistant line, unless another boundary line is justified by 'special circumstances'. The overlapping area covers about 155,000 km^2, most of which is permanently free of ice and hence accessible for all military and economic purposes.

The sector theory as a justification for sovereign rights was advocated in the past by Canada and the Soviet Union.[19] It may apply to land territory[20] and islands but certainly not to ocean space,[21] otherwise the whole Arctic Ocean would be placed outside the framework of international law of the seas and the Soviet Union would appropriate 43 per cent or 6.8 million km^2 of the Arctic Ocean.[22] The Soviet Union itself has declared a continental shelf zone in 1968,[23] a 200-nautical miles (nm) fishing zone in 1976[24] and a 200 nm EEZ in 1984[25] for its own offshore waters including all of the seas north of the Soviet Arctic coast. These declarations of sovereign rights over fish and other resources would have been redundant if the Soviet Union really had stuck to the sector principle as being applicable to ocean spaces. Moreover, one may question the sector theory in the light of the LOS Conference, where various modes of extending coastal-state jurisdiction have been worked out, but none of these proposals gives any further credit to the sector theory.[26] Consequently, sector claims in the Arctic can no longer be taken seriously. Norway has rejected the sectoral line arguments on the grounds that its use is highly controversial in international law. The Soviet Union, as mentioned earlier, maintains the sector line at least as an 'academic' delimitation *criterion*, not as a *principle*.

Furthermore, the Barents Sea delimitation problem is complicated by the legal status of Svalbard (Spitsbergen). This archipelago, comprising 62,000 km^2 of islands, is situated 355 nm north of the North Cape. The Svalbard regime applies to all islands in a rectangular area (the so-called 'Spitsbergen Net') between 10°E and 35°E and between 74°N and 81°N, including the strategically interesting Bear Island, which is situated half way between Norway and Svalbard. The Soviet sector claim, although following in principle 32° 04' 35" E,[27] coincides with the eastern limits of the Spitsbergen Treaty, thereby respecting Norway's sovereignty (see Figure 3.1).[28] The unique legal status of Svalbard has been defined in the Svalbard Treaty of 1920,[29] which granted Norway 'full and absolute sovereignty...over the archipelago of Spitsbergen',[30] subject to the condition that the economic resources of the islands, including fishing and mining, are free, 'both on land and in the territorial sea',[31] for all of the 41 signatory states on a footing of absolute equality, and that the islands are demilitarised and may 'never be used for warlike purposes'.[32] From a legal point of view it is not a condominium but rather Norwegian sovereignty limited by certain rights of third parties. Apart from Norway only the Soviet Union makes use of their rights under the Svalbard Treaty by operating coal mines on the islands with the help of about 2,000 Soviet personnel.[33] Other nations like France and the Federal Republic of Germany have restricted their Svalbard activities to fishing and to scientific research.

It seems clear that in the context of 1920 the contracting parties of the Spitsbergen Treaty did not encompass concepts such as the continental shelf and EEZ.[34]

In the Norwegian view the specific purposes and regulations of the Svalbard Treaty apply only to the islands proper, including the 4 nm territorial sea.[35] The continental shelf and other maritime zones beyond the 4 nm limit are subject to unrestricted Norwegian sovereignty under international law. In June 1977, Norway proclaimed a non-discriminatory fisheries protection zone of 200 nm for Svalbard[36] thus stopping short of proclaiming a 200-nm EEZ. Besides, Norway regards the shelf areas around Svalbard as part of an uninterrupted North Norwegian shelf, although the 'Barents Trough' (south of Bear Island) with water depths of over 500 m may be regarded as a natural separation of the North Norwegian and Svalbard shelves. By claiming the shelf in this way Norway avoids the interpretative questions associated with the 1920 Treaty.[37] USA, USSR, Denmark and the United Kingdom have raised reservations against these Norwegian shelf arguments, maintaining that a treaty which grants non-discriminatory economic rights must be extended to cover an EEZ and a continental shelf.

In particular the Soviet Union advocates that Svalbard has its own continental shelf where the 'economic condominium' of the Svalbard Treaty applies. If this were the case, all signatory states including the USA and many NATO states would be entitled to exploit not only the terrestrial resources, but also the rich offshore areas in the Svalbard area. At the same time the demilitarisation clause of the Svalbard Treaty would apply to the continental shelf and consequently some naval activities would be illegal there.

The legal regime of islands, in the light of the 1958 Geneva Shelf Convention and even more so in the light of Article 121 of the new

LOS Convention, clearly institutes full maritime zones for all islands. Consequently, Norway will, in the long run, restrict itself to the legal position that the Svalbard Treaty applies only to the islands proper, while Norway's 'full and absolute sovereignty' over the islands will grant unrestricted jurisdiction over the adjacent maritime zones. Nevertheless, one may ask whether continental shelf areas adjacent to uniquely regulated land areas are subject to the same regime as the land areas,[38] a question which would also be relevant to Antarctica.

In view of these uncertainties, Norwegian authors rightly stressed that a 'free-for-all' in the maritime zones of Svalbard would create conditions of tension and disorder[39] in a very sensitive part of the world. A new Svalbard conference or an arbitration procedure, as alternatives, would bring in outside factors and legitimate interests of the other signatory states. In view of this, the Soviet position may be primarily of a tactical nature.

> The hypothesis is that the Soviets are anxious to obtain a package deal in which the Soviet Union would waive its resistance to the Norwegian continental shelf view in return for Norway accepting a dividing line closely approximating the sector line.[40]

A political solution, if any, seems to be the only way out of this dilemma, which involves the whole spectrum of delimitation problems of adjacent and opposite areas, as well as special circumstances.

Regardless of this speculation, the introduction of the 200-nm fishing and economic zones has led to an interim arrangement over a so-called 'Grey Zone',[41] as in force from March 1978 in an area of 67,420 km^2 north of the Norwegian–Soviet land border. The Grey Zone agreement has to be renewed annually in order to remain in force, as happened when it was renewed until 1 July 1988. It provides for parallel Soviet and Norwegian fishery jurisdiction in the entire zone on an interim basis without prejudicing the lateral delimitation problems for the EEZ and the continental shelf. More than 90 per cent of the Grey Zone lies west of the Norwegian-claimed median line, and much of it even west of the Soviet-claimed sector line. The geographical imbalance of the agreement has caused widespread uneasiness in Norway. This Grey Zone agreement of a parallel Soviet–Norwegian regime is probably very attractive to the Soviet Union. Its underlying concept of exclusive economic co-operation or bilateralisation could be the aim of the Soviet maritime policy for the whole area, including Spitsbergen. As mentioned above, there is already a close Norwegian–Soviet co-operation in the planning stage for hydrocarbon exploration in the Barents Sea. However, no real co-operation can be expected in an area in which military pressure and unresolved legal issues prevail.

The latest incident in the area occurred in July 1985 when the Norwegian research vessel '*Malene Østervold*', exploring on behalf of the Norwegian Oil Directorate 50 km west of the Soviet-claimed sector line, lost part of its underwater equipment due to the intervention of a Soviet frigate.

SOVIET MARITIME CLAIMS AND OPTIONS

A common characteristic of all Arctic maritime borders of the Soviet Union is their lack of precision. There are no known Soviet maps published and available which show its territorial sea or other maritime borders. Moreover, there are no known straight baselines.[42] The USSR Declaration 4604 informs us that the Council of Ministers Decree of 7 February 1984 approved a list of co-ordinates which define straight baselines off the coasts of the Pacific Ocean, the Sea of Japan, the Sea of Okhotsk, and the Bering Sea; but no straight baselines were listed for the Arctic coastline.[43] Elizabeth Young[44] reports that in July 1983, the Soviet government informed Western ambassadors in Moscow that work was in hand to establish straight baselines in the Arctic Seas and that details would be published on completion of the exercise but there were no results known so far. In fact the determination of precise baselines in the Arctic remains a difficult job for various reasons (absence of adequate charts, the unreliability of Mercator projections in the far north, the difficulty in identifying base points in ice-covered areas, etc.).[45] It will require an enormous amount of costly surveying before the Arctic limits of the Soviet Union could be precisely fixed. Moreover, it may well be the Soviet Union's intention not to be too explicit on the delimitation issues at this time.

In addition to the 12-nm territorial sea,[46] the Soviet Union declared a provisional 200-nm fishery zone on 10 December 1976, to become effective as from 1 March 1977.[47] As recorded in the preamble to this declaration, the Soviet government reacted with reluctance to the 200-nm declarations of other states, e.g. the United States and the EEC states. The provisional 200-nm zone extends to a distance of up to 200 nm from the baselines from which the territorial waters of the USSR are measured. The marine areas affected include all of the Arctic waters north of the Eurasian coast and the areas around islands belonging to the USSR, but excluding the Barents Sea which became the subject of a separate declaration in May 1977.[48] The fishing zone declaration lacks an answer as to the nature of the western boundary in the Barents Sea.

In the Bering Strait and Chukchi Sea, the eastern boundary of the fishing zone is the Russo-American convention line of March 1867,[49] i.e. longitude 168° 49' 30" W, which the US also observed in its national legislation although disagreeing with the sector theory in principle.[50] Regarding this delimitation line the official position of the US is that the convention line of 1867 does not extend into the polar regions beyond 72° N, on the grounds that the parties of 1867 could not have had in mind anything else but land possessions, since the continental shelf concept was still unknown.[51] As in the case of the Spitsbergen Treaty, historic treaties — conceived for specific purposes — should be interpreted in the historic context and not misunderstood as a 'living instrument'.[52]

In January 1984, the Soviet Union notified the United States that it was prepared to resume discussions on demarcation of the maritime boundary in the Bering Strait. According to news reports,[53] negotiations took place later in 1984 with results so far unknown. In addition, under unknown circumstances the US vessel '*Frieda K*' and her crew were stopped and detained by Soviet forces in this area on the grounds of an 'illegal violation' of the Soviet border.[54] In the absence of any bilateral agreement, the Soviet–US border line in the Chukchi Sea and marine areas north of it, remains unresolved.

The Soviet Union exercises sovereign rights over the continental shelf according to a decree of 1968.[55] This decree applies to submarine areas adjacent to the coast, or to islands, to a depth of 200 m or, beyond that limit, to where the depth of the superjacent waters admits the exploitation of natural resources.

Both Soviet decrees on the fishery zone and on the continental shelf zone have been superseded most recently by the decree establishing a 200-nm exclusive economic zone, to become effective as from 1 March 1984.[56] The declaration is more or less in accordance with Parts V and VI (EEZ, continental shelf) of the new LOS Convention. A preliminary analysis shows, however, that in several instances the sovereign rights, jurisdictions and other rights are interpreted in an extensive way, going beyond what has been agreed upon in the Convention. When it comes to delimitation, the decree refers to the 200 nm and, in the case of lateral limits, to agreements with neighbouring states or states opposite on the basis of international law with the aim of achieving a just decision. Consequently, this decree introduces no new elements to the Bering Strait and Barents Sea limits.

Two major uncertainties remain regarding the extent of Soviet Arctic waters. One refers to the legal status of certain areas occasionally claimed as 'historic bays' or 'historic waters' to be part of internal waters. The Kara, Laptev, East Siberian and Chukchi Seas should be mentioned here. The international law and practice is rather vague with respect to these types of waters.[57] Historic bays are mentioned in Article 10 para. 6, but the Soviet Union will find little support in having these vast seas respected as historic bays in the sense of this Article.

Another issue that remains to be determined is the ultimate outer continental shelf limit in the Arctic, provided the Soviet Union or any other Arctic state decides to make use of the relevant provisions of Article 76. In the light of the new LOS Convention the 'traditional' shelf claims have to be reconsidered and adapted to the unique characteristics of the Eurasian shelf, which is one of the broadest shelves in the world, extending in many areas far beyond the 200 nm with a maximum of 1,700 km.[58]

The average extent of the Eurasian shelves may be in the order of 300 to 400 nm ending up in the four Arctic basins or abyssal plains with water depths of 3,000 to 4,000 m:

- the Nansen Basin
- the Amundsen Basin
- the Makarov Basin
- the Canada Basin/Beaufort Sea.

The four basins are separated from each other by three mid-ocean ridges with water depths of 1,000 to 2,000 m:

- the Arctic Mid-ocean Ridge
- the Lomonosov Ridge
- the Alpha Ridge/Mendeleyev Ridge.

The three parallel ridges cut across the Arctic Ocean connecting the East Siberian continental shelf with the shelves north of Greenland. If these ridges were in fact natural components of the shelves there might be a chance of hydrocarbon deposits.

Against this background of geological structures, the limits of the continental shelf, as laid down in Article 76, deserve special attention. By the terms of this provision the outer continental shelf beyond the 200-nm limit will 'normally' be determined in accordance with the so-called 'Irish formula' of Article 76 para. 4, based on geomorphological criteria such as thickness of sediments, but with an outer limit of either 350 nm from the baselines or 100 nm from the 2,500 m isobath. Given these two alternatives the coastal state may choose the most convenient border line. However, special provisions of Article 76 para. 6 apply to 'submarine ridges' where the limits shall not exceed 350 nm, unless these submarine elevations are natural components of the continental margin, such as its plateaux, rises, caps, banks and spurs.

The repercussions of these elaborate definitions are not yet fully apparent. But one may anticipate that they allow for a maximum of nationaliastion of the Arctic shelves far beyond the 200 nm, especially on the East Siberian shelf and on the neighbouring Alaskan shelf including the Chukchi plateau.

The details of this long-contested shelf limitation were introduced into the new Convention in 1980,[59] promoted by a few states with extraordinarily broad shelves ('the margineers') including the United States[60] and the Soviet Union,[61] thereby overriding objections by geographically disadvantaged states.[62] It is obvious that the outer shelf regime, especially its outer limits in the case of ridges, banks and plateaux, has been designed to accommodate the territorial interests of the two superpowers and a few others at the expense of the common heritage. The prospect for Arctic expansion might have been the Soviet motive to promote this idea from the beginning of the 3rd UN Conference on the Law of the Sea,[63] while the USA is profiting from this concept in the North-east Pacific Ocean where another system of mid-ocean ridges qualifies for nationalisation.

An interesting test case of the scope of Article 76 para. 6 might be the Chukchi plateau situated to the north of Alaska and the Bering Strait. Will the Chukchi plateau be part of the continental shelf at all, and if so, will it be under Soviet or US jurisdiction?[64] In the absence of precise shelf limits, all Arctic states will have the legal opportunity to claim outer shelves beyond 200 nm — with more delimitation problems to follow.

NOTES

1. W.E. Butler, *The Northeast Arctic Passage* (Sijthoff and Noordhoff, Alphen aan den Rijn, 1978); see also USSR Statute of the Administration of the Northern Sea Route of Sept 1971, in H. Lay, R. Churchill and M. Nordquist (eds), *New directions in the Law of the Sea*, Vol. II (London, 1973), p. 710.

2. *Journal of Commerce*, 18 April 1983.

3. For example, MIZEX — Marginal Ice Zone Experiment 1983/84, to be conducted by seven European states including the Federal Republic of Germany (polar research vessel *Polarstern*), Canada and USA; another project (Fram Strait Project) is to study the flows of water and ice into and out of the Arctic Ocean.

4. D. Pharand, 'The implications of changes in the Law of the Sea for the "North American" Arctic Ocean' in J.K. Gamble (ed.), *Law of the Sea: neglected issues* (Law of the Sea Institute, University of Hawaii, 1979), p. 187 et seq.

5. *Oil and Gas Journal*, 20 August 1984.

6. F. Sollie, *The economic and strategic importance of the Norwegian and Barents Sea* (Fridtjof Nansen Foundation, Oslo, 1978), p. 3; Swedish sources estimate that the Soviet offshore oil and gas development requires some US $ 80 billion.

7. Bo Johnson Theutenberg, *The evolution of the Law of the Sea: a study of resources and strategy with special regard to the polar regions* (Tycooly, Dublin, 1984), pp. 26–35. As regards the legality of military installations the 'Treaty on the Prohibition of the Emplacement of Nuclear Weapons' of 1971 (ST/LEG/SER.B/19 p. 455) has to be observed banning (only) the stationing of weapons of mass destruction beyond the 12 nm limit.

8. Elmar Rauch, *Politische Konsequenzen und Möglichkeiten der Seerechtsentwicklung aus der Sicht der UdSSR, Berichte des Bundesinstituts für ostwissenschaftliche und internationale Studien*, Vol. IV (Köln, 26-1979), p. 38.

9. International Institute for Strategic Studies, *The military balance 1986/87* (IISS, London, 1986); *Frankfurter Allgemeine Zeitung*, 8 August 1984.

10. Elmar Rauch, *Politische Konsequenzen*, Vol. II, 47-1977, p. 15.

11. S.M. Olenicoff, *Territorial waters in the Arctic: the Soviet position* (The Rand Corporation, Santa Monica, July 1972, R-908-ARPA). See also Akademie der Wissenschaften der UdSSR, *Modernes Seevölkerrecht*, Baden-Baden, 1978, p. 214–19.

12. W. Østreng, 'The continental shelf — issues in the eastern Arctic Ocean' in Gamble (ed.), *Law of the Sea*, p. 169; John Norton Moore, *American Journal of International Law* (AJIL), vol. 74 (1980), p. 232; *Arctic Ocean issues in the 1980s*, p. 3 and 12.

13. F. Sollie, 'The polar seas: issues not dealt with in the Convention, reasons and problems', *Law of the Sea Institute Proceedings*, vol. 17 (1984), p. 654.

14. Regarding the status of Wrangel Island, see *AJIL*, vol. 72 (1978), p. 894 and J.P.A. Bernhardt, 'Sovereignty in Antarctica', *California Western Journal of International Law*, 1975, p. 334; regarding Hans Island (Kennedy Channel) it is interesting to note that the Danish–Canadian shelf delimitation agreement leaves this contested island out of consideration.

15. *LOS Bulletin*, no. 1 (September 1983), p. 30.

16. The only settled problem refers to the lateral border line in the Varangerfjord Agreement of 15 February 1957 and Protocol of 29 November 1957, United Nations Treaty Series (UNTS), vol. 312, no. 289; see also *Atlante Dei Confini Sottomarini* (Milan, 1979), p. 3.

17. W. Østreng, 'The continental shelf', p. 166; Rauch, *Politische*

Konsequenzen, vol. I, 36-1977, pp. 54–61; C.A. Fleischer, 'The Northern Waters and new maritime zones', *German Yearbook of International Law, GYIL*, 22 (1979), pp. 100–18.

18. Norway and the Soviet Union have ratified this Convention.

19. The Soviet Sector Decree of 1926 claims sovereignty over islands in the sector between 32° 04' 35" E and 168° 49' 30" W, but not over ocean space and/or continental shelves. Regarding early Canadian statements of 1907 and 1925 on the sector theory, see Peter Bernhardt, 'Sovereignty in Antarctica', *California Western Journal of International Law*, 1975, pp. 332–8; Rauch, *Politische Konsequenzen*, vol. I, 36-1977, p. 55; see also Theutenberg, *The evolution of the Law of the Sea*, p. 37.

20. For a general rejection of sector claims to all land and sea areas, see Bernhardt, 'Sovereignty in Antarctica', pp. 332–8.

21. Victor Böhmert, *Sektorentheorie, Wörterbuch des Völkerrechts*, vol. III, p. 248; P.C. Barabolya *et al.*, *Manual of international maritime law*, (Ministry of Defence, Moscow, 1966), vol. II, pp. 269–73.

22. See Rauch, *Politische Konsequenzen*, vol. I, 36-1977, p. 69 with reference to Soviet sources; and Akademie der Wissenschaften der UdSSR, *Modernes Völkerrecht*, (Baden Baden, 1978), p. 216.

23. Soviet Decree of 6 February 1968 on the Continental Shelf of the USSR, English text in ST/LEG/SER.B/15, p. 441.

24. English text in A/Conf 62/53 of 15 February 1977; see also W.E. Butler 'Soviet fishery jurisdiction in the Arctic' *Polar Record*, vol. 18, no. 117 (1977), pp. 575–9.

25. Soviet Decree of 28 February 1984.

26. *Arctic Ocean issues*, pp. 15 and 47.

27. See explanation in note 19.

28. Robert W. Smith, 'National claims and the geography of the Arctic' paper presented at the 18th LOS Institute meeting, San Francisco, September 1984, p. 27.

29. Treaty Regulating the Status of Spitsbergen of 3 February, 1920, *Reichsgesetzblatt* RGB1 1925 II, p. 763; see also Ingo von Münch, 'Spitzbergen', *Wörterbuch des Völkerrechts*, vol. III, p. 300; Rauch, *Politische Konsequenzen*, vol. I, 36-1977, p. 56; R. Platzöder, *Politische Konzeptionen zur Neuordnung des Meeresvölkerrechts* (Ebenhausen, 1976), p. 135–48.

30. Spitsbergen Treaty Article 1.

31. Spitsbergen Treaty Articles 2 and 3.

32. Spitsbergen Treaty Article 9.

33. Rauch, *Politische Konsequenzen*, vol. IV, 26-1979, p. 40.

34. P.A. Bernhardt, 'Spitzbergen: jurisdictional friction over unexploited oil reserves', *California Western Journal of International Law*, Winter 1974, pp. 110–15.

35. Claimed since 1970 for Spitsbergen.

36. As note 27, the Soviet Union rejected the Norwegian declaration — see Rauch, *Politische Konsequenzen*, vol. IV, 16-1979, p. 42.

37. Smith, 'National claims', p. 37.

38. This view is advocated by Platzöder, *Politische Konzeptionen*, p. 144; see also Rauch, *Politische Konsequenzen*, vol. I, 36-1977, p. 58.

39. Østreng, 'The continental shelf', p. 175.

40. Ibid.; see also Platzöder, *Politische Konzeptionen*, p. 146.

41. German text in Rauch, *Politische Konsequenzen*, vol. II, 47-1977, p. 36, with further details of the Norwegian–Soviet negotiations, ibid., pp.14–19.

42. Smith, 'National claims', p. 26.

43. Ibid., and *Arctic Ocean issues*, pp.15 and 47.

44. Elizabeth Young, 'Soviet Arctic shipping', *Cultures et Sociétés de l'Est* (Institut d'Etudes Slaves, Paris, 1984), p. 328.

45. Smith, 'National claims', p. 17.

46. Law on the State Boundary of the USSR in *International Law Material*, 1983, 1055–76.

47. See references in note 24.

48. Barents Sea fishery zone, effective as from 25 May 1977; Rauch, *Politische Konsequenzen*, vol. II, 47-1977, p. 13.

49. Butler, 'Soviet fishery jurisdiction', pp. 575–9, also providing a map.

50. Mark Feldman and David Colson, 'The maritime boundaries of the United States', *AJIL* vol. 75, (1981), p. 751.

51. For further details and analysis of the 1867 Convention ceding Alaska, see Bernhardt, 'Sovereignty in Antarctica', p. 333 and US Department of State, Bureau of Intelligence and Research, *US–Russia Convention Line of 1867* (Washington DC, 1975), p. 3; Pharand, 'The implications of changes', p. 184.

52. Bernhardt, 'Spitzbergen', p. 113.

53. *Frankfurter Allgemeine Zeitung*, 27 July 1984.

54. *Frankfurter Allgemeine Zeitung*, 19 September 1984.

55. Soviet Decree (6/2/68) on the Continental Shelf of the USSR.

56. *Decree of the Presidium of the Supreme Soviet Establishing a 200-mile Economic Zone of February 28, 1984*, unofficial German translation.

57. For details see Smith, 'National claims' pp. 25–33; U. Jenisch, *Das Recht zur Vornahme militärischer Übungen und Versuche in Friedenszeiten*, (Forschungsstelle für Völkerrecht u. ausländisches öffentliches Recht der Universität Hamburg, 1970), p. 36–42.

58. Günter Hepper, 'Der Festlandsockel', in *Seewirtschaft* (Berlin/Ost, 1983), p. 479.

59. ICNT/Rev. 2 of 11 April 1980; cf. the so-called Aguilar proposal A/Conf 62/L 51 of 29 March 1980 and NG 6/21 of 18 March 1980.

60. US proposal NG 6/11 of 13 August 1979.

61. USSR proposal NG 6/8 of 18 April 1979 and NG 6/9 of 9 August 1979.

62. For example, Arab states and Austria, cf. NG 6/2 of 11 May 1978; Singapore NG 6/17.

63. As regards early USSR proposals for a 500-m depth-line, see Rauch, *Politische Konsequenzen*, vol. III, 22-1978, p. 12 and vol. IV, 26-1979, p. 28.

64. Cf. statement by US Ambassador Richardson, A/Conf 62/128 of 25 April 1980, p. 39.

5

Soviet Fisheries and Offshore Exploration in the Barents Sea

David Scrivener

FISHERIES IN THE BARENTS SEA

The common need effectively to manage the rich yet vulnerable fish stocks of the Barents Sea — particularly joint stocks of cod — came to be felt more acutely in the 1970s in both Norway and the USSR. The declaration of a Norwegian Exclusive Economic Zone (EEZ) and Soviet Fisheries Protection Zone (FPZ) in late 1976 brought the delicate prospect of overlapping fisheries jurisdiction claims and raised Norwegian concern over potential conflicts in inspection of third-country vessels in the water out to 200 nautical miles (nm) above the disputed shelf. The two governments had agreed earlier in the year on mutual access, setting of total catches and allocation of quotas in their respective national zones through the Norwegian–Soviet Fisheries Commission. Negotiations stemming from a Norwegian initiative led to the temporary bilateral fisheries management agreement of January 1978 which entered into force in March 1978. The 'Grey Zone' agreement covered an area out to 200 nm of 67,420 km^2 (26,000 $miles^2$), of which 41,000 km^2 lay above the disputed shelf. The Grey Zone embraced 8,900 $miles^2$ to the west of the Soviet sector line and 1,200 $miles^2$ to the east of the Norwegian median line, north-east of Rybachii Peninsula. Total Allowable Catches (TACs) and quotas were to be agreed with the bilateral Fisheries Commission. Joint policing was avoided by allowing each state exclusive jurisdiction over its own vessels and those of third countries it licensed to operate in the Grey Zone. The arrangement has worked well on a practical level and has been renewed annually with relatively few difficulties. Both states clearly have a major stake in enhancing Barents Sea fisheries. Joint fisheries management and co-operation in protecting species previously subject to rapid depletion facilitated a marked resurgence of such stocks as cod, haddock and herring. Joint fisheries research expeditions in the Barents began in the early 1960s and have proved highly useful to both sides.

Table 5.1: Main Soviet nominal catches (by species) in the North-east Atlantic Fishing Area, 1982

Species	Fishing Area	Total Soviet Nominal Catch (tonnes)	Proportion of total species catch in the area (%)	Proportion of total Soviet species catch in NE Atlantic (%)
CAPELIN	Barents Sea	428,082	50.0	72.0
	Svalbard-Bear Island	12,815	20.0	2.0
	Norwegian Sea	155,585	8.0	26.0
BLUE WHITING	Svalbard-Bear Island	59	100.0	0.0
	Norwegian Sea	103,711	94.0	58.6
	Faroe Plateau	71,371	69.0	40.0
REDFISHES	Norwegian Sea	63,125	79.6	36.5
	Svalbard-Bear Island	47,935	96.0	28.0
	North-east Greenland	20,217	97.0	11.7
	Barents Sea	1,750	68.0	1.0
COD	Barents Sea	17,669	18.4	14.0
	Svalbard-Bear Island	16,042	51.7	12.6
	Norwegian Sea	6,600	2.4	5.0
HERRING	Barents Sea	889	100.0	0.9
POLAR COD	Barents Sea	90,084	100.0	99.8
	Svalbard-Bear Island	148	67.0	0.2

The figures in Tables 5.1 and 5.2 were computed from data in the ICES *Bulletin Statistique des Peches Maritimes*, vols. 66 and 67.

In 1982 the Barents Sea yielded 5.9 per cent of the world-wide Soviet fish catch and 21.9 per cent of the Norwegian. The USSR took 51 per cent of the total nominal catch in the Barents (59 per cent in 1981) while Norway took 47.5 per cent (40 per cent in 1981). The largest Soviet catches were of capelin, cod and shrimp (see Table 5.1).

Third-country catches in the Barents declined considerably with the introduction of 200 nm Soviet and Norwegian fisheries and economic zones in the mid-1970s. In 1982 the Faroes took 11,500 tonnes of cod and capelin amounting to just 1 per cent of the total nominal fish catch in the Barents. The UK took 3,700 tonnes of cod (0.3 per cent) while France and the GDR caught 615 and 136 tonnes of fish respectively. The previous year had also seen catches of a few hundred tonnes by the Federal Republic of Germany (FRG), Poland and Spain.[1]

The establishmnent and regular renewal of the Grey Zone arrangement, in addition to meeting a shared Soviet concern to enhance fisheries management works in favour of the Soviet position on shelf delimitation. The inclusion of such a large area west of the sector line is probably viewed as a precedent for future Norwegian concessions on economic zone and shelf delimitation and may tempt Moscow to regard the western boundary of the Grey Zone as a *de facto* boundary.[2] Moscow may take the agreement to indicate a Norwegian readiness eventually to compromise on its insistence on the median line. Molodtsov cites the agreement as a good example of the utility of 'temporary measures' in avoiding tension and facilitating resolution of outstanding boundary disputes. Though allowing that such measures are by nature temporary and do not prejudice the final zone and shelf delimitation, he depicts the agreement's existence as a further argument undermining the equidistance principle.[3] The area embraced by the Grey Zone was determined according to fisheries management rather than legal criteria though the Soviets make no effort to stress this. The western extreme of the zone mostly corresponds with the ICES (International Council for the Exploration of the Sea) 30° E statistical boundary dividing the Barents (I) and Norwegian Seas (IIa) fishing areas.

The fact that the agreement covered a greater area to the west of the sector line than to the east of the median line was one of the main criticisms made by opponents of the accord in Norway. Its success from the fisheries conservation viewpoint prompted some Norwegian circles to question the need to continue the arrangement on purely conservation criteria. Avoidance of tension with the USSR over enforcement of fisheries regulations with respect to third-country vessels in the disputed area remains the major motive behind Oslo's willingness to maintain the Grey Zone arrangement.

In the absence of agreement on shelf and zone delimitation, the USSR may use the Grey Zone accord to encourage consideration of joint management or consultation processes in other functional spheres (e.g. oil exploitation) with the longer-term aspiration of realising some form of exclusive condominium arrangement. Moscow has tried to treat the area of the Baltic disputed with Sweden as closed to third-country fishing, on occasion boarding Danish and West German vessels and presenting forms printed in Russian and Swedish for completion.[4] Similar incidents have been very rare in the Barents Grey Zone; the Soviets boarded a British trawler in Spring 1978 but subsequently explained this as an error on the Soviet captain's part. The terms of the agreement avoided instituting joint enforcement of fisheries regulations through co-ordinated inspection and thus maintained a regime of separate jurisdictions. Yet the few Soviet commentaries on the agreement that exist leave it unclear whether each state's jurisdiction over third-country vessels is limited to those licensed by it.[5]

The Grey Zone arrangement met the Soviet interest in conserving joint fisheries stocks in the Barents Sea while being open to interpretation as a precedent for shelf and zone delimitation with Norway. It also shares some similarities with and implications for Soviet policy on fisheries jurisdiction in the Svalbard–Bear Island area. Finally, it may contribute to Norwegian–Soviet offshore energy co-operation in a manner conducive to the Soviet position on shelf delimitation.

SVALBARD FISHERIES

The Soviets questioned the legal basis of Norway's introduction of a 200 nm FPZ in the archipelago in June 1977. In this case, the USSR was joined by Spain with its growing interest in Barents Sea fisheries. The Soviets have largely complied with the regulations, conforming with required net sizes, observing off-limits areas and submitting to net inspections by Norwegian fisheries protection vessels. Norway faced enforcement problems mainly with Spanish rather than Soviet-bloc fishing vessels. Soviet catches find their way into ICES statistics. To report catch sizes directly to Norway was probably seen as implicit recognition of Norway's right to establish a zone and a precedent for other jurisdictional issues on and around Svalbard. Soviet compliance nonetheless marks the common interest in conserving the vulnerable fish stocks in the Svalbard–Bear Island area, which are of some importance to the Soviet fisheries industry.

In 1982 the area yielded 2.5 per cent of the world-wide Soviet fish catch and nearly 24 per cent of the Norwegian. The USSR took 28 per cent of the total nominal catch around Svalbard (20.7 per cent in 1981), while Norway took 66 per cent (76 per cent in 1981). The largest Soviet catches were of capelin, redfishes, and Greenland halibut in 1981. In 1982, with the resurgence of cod stocks, capelin, redfishes and cod were the three top Soviet catches (see Table 5.2). The two states together accounted for nearly 97 per cent of the total catch in the area in 1981 and for 94 per cent of the 1982 catch which had almost doubled in size after the roughly 50 per cent reduction between 1979 and 1981. In 1982 Denmark reappeared on the scene, taking 2 per cent of the total catch, while the Faroes took 1.9 per cent and East Germany (GDR) 0.3 per cent. In taking nearly half the cod in 1982 Spain accounted for 1.6 per cent of the total nominal catch. Eight British vessels fished in the area in 1982, landing 34 tonnes of cod in total, compared with three boats landing 16 tonnes in 1981.[6]

That the fisheries zone has so far been non-discriminatory could be interpreted in Moscow as confirming the applicability of the equal access and treatment provisions in the Svalbard Treaty beyond the archipelago's territorial waters. Furthermore, Norway consulted the USSR and the European Community before introducing the detailed fisheries regulations in 1978. Moscow can make common cause with Spain — and now conceivably the European Commission — over fisheries jurisdiction around Svalbard.[7]

Soviet hopes of access to Greenlandic fisheries received a boost in 1982 when Soviet vessels took 22,400 tonnes (mainly redfishes) in the North-east Greenland fishing area. This amounted to 97 per cent of the total nominal catch in the area, while Poland's redfish take accounted for 2.5 per cent. Moscow's current effort to establish fisheries co-operation with the Greenlandic Home Rule government is a further exploitation of the fisheries-security interconnections in NATO intra-alliance politics. However unclear Moscow's designs are, the USSR could be encouraged by complications surrounding future delimitation of the continental shelf between Greenland and Svalbard.[8]

OFFSHORE EXPLORATION IN THE BARENTS SEA

In the late 1970s exploration on the Norwegian continental shelf above 62° N began to expand quite rapidly. The Haltenbanken and Tromsøflaket areas have already given very encouraging results. In 1985-6 new blocks were to be opened up for exploratory drilling including several in the Bjornøya (Bear Island) and Finnmark areas in the search for oil and to encourage regional economic development in

Table 5.2: Percentage take of total nominal catch of the most important species in the Svalbard-Bear Island Fishing Area, 1981 and 1982
Total nominal catch: 1981 - 436,589 tonnes; 1982 - 903,400 tonnes

Species	Total catch 1981 1982	Norway	USSR	Faroes	Spain	GDR
CAPELIN	382,273 773,389	85.7 75.8	12.3 20.0	1.76 1.76		
REDFISHES	27,023 49,883	0.35	89.9 96.1		1.14 0.14	8.9 3.4
COD	2,032 31,034	4.4 0.4	73.8 51.7	12.8 0.5	46.8	8.0 0.5
UNSORTED	4,478 11,980		96.0 100.0		4.0	
GREENLAND HALIBUT	8,504 11,029	5.8 1.9	78.6 88.3			15.8 9.8
CATFISH	872 4,869		99.9 99.9			
SHRIMP	7,642 17,401	49.0 59.0	42.0 19.2	8.5 19.2		
VARIOUS PLECTO-NEURIFORMS	2,032 3,098		99.5 99.94			0.5 0.06

northern Norway. Esso and Elf have already joined Statoil and Norsk Hydro in setting up offices in Harstad and Hammerfest and the Norwegian government is considering development in Kirkenes as an offshore supply base for the future opening up of the Barents Sea. Oslo maintains a certain anxiety to attract foreign participation on the Norwegian Barents shelf in order to secure a counterweight to what will probably be an intensified Soviet offshore programme in the next decade.

The Soviet Union began extensive seismic surveys in the Barents Sea in 1978, conducting three comprehensive geological and geophysical expeditions in the southern part in 1980-2 aboard the *Professor Shtokman* research ship. Three theoretically promising areas were identified west of Novaya Zemlya and in the south-eastern Barents Sea region. In late Spring 1982 the Finnish-built ice-class drill ship *Valentin Shashin* — capable of drilling in up to 300 metres (m) of water — spudded its first well on the Murmansk High structure (70° N, 42 E), with help from divers aboard a Finnish-built research ship. In 1983 two ships operated in the Barents, finishing the earlier well and opening another nearby. A third well was spudded by the *Valentin Shashin* on a structure near the disputed shelf area. Wells were also spudded in quite shallow water in the Pechora Sea. In late 1984 Soviet officials began hinting at a major oil and gas find in the central portion of the Barents. The Finnish firm Rauma Repola delivered the first two of three Arctic-conditioned Gusto-type jack-up rigs to the USSR, one of which was certainly destined for work in the Barents Sea. In early 1985 a further order was placed in Finland for legs and elevation systems for two Arctic jack-up rigs being built by the Soviets at Vyborg which will weigh 14,500 metric tons. The order, worth $46 million, was to be completed by 1987. This followed the news of an order placed with the Brazilian firm Nobara for four jack-ups and options on a further six, although these would probably not be capable of operating in Arctic waters. In May 1985 the Shelf-4 Arctic-capable semi-sub built by the Soviets at Vyborg was delivered to the waters off Kolguyev Island in the south-east Barents, capable of drilling in water depths of up to 200 m. It carries saturation diving equipment provided by Seaforth Maritime of Aberdeen. Exploration wells were drilled on Kolguyev Island itself in the early 1970s. The rig will probably begin drilling in a field near the mouth of the Pechora River. Finnish and Norwegian firms have sold over 15 offshore vessels to the USSR, at least one of which was working in the Barents in 1983.

Soviet offshore energy activities accelerated in the mid-late 1970s in a number of areas. The Soviets have acquired most of their relatively limited experience in exploration and production in the Caspian Sea where three jack-ups and one semi-sub were at work by 1980. In that year development of the shelf series of semi-subs began at Astrakhan and Baku: these are capable of drilling in up to 200 m of water and two are at present operational. The bulk of production comes from fixed platforms and the new French-built Karadag yard will soon produce larger fixed structures able to drill deeper than those now operating in up to 110 m of water on the '28 April' deposit. A special ship to tow and help install Karadag platforms arrived from West Germany in 1983. Moscow no doubt views its effort in the Caspian as invaluable in amassing offshore operating experience and contributing to the improvement of the Soviet technology and equipment base.[9]

In 1975 the USSR, Poland and the GDR formed the Petrobaltic consortium jointly to explore their Baltic continental shelves with its offices in Gdansk. Exploration costs were to be financed equally and 50 per cent of any oil found would go to the state on whose shelf it is found, the rest being divided between the other partners. Petrobaltic acquired a jack-up rig from Holland in 1979 able to drill in depths of 90 m which is jointly manned. The Soviet Union made its first interesting find in the Baltic off Cape Taran where a fixed platform began operating in 30 m of water in early 1984.

Joint exploration of the Sakhalin shelf in the Sea of Okhotsk began with Japan in 1977 using two mobile rigs. A 1975 agreement provided for Japan to supply equipment and ships for prospecting and the necessary credits. Two major deposits have so far been opened — Odoptu and Chaivo — and development plans are now being drawn up for production in the early 1990s, including several fixed platforms and gas pipelines to the Sakhalin mainland. By mid-1985 a Soviet Sivash jack-up had drilled its second hole on the Bulgarian shelf in the Black Sea.

The Soviets have so far had only very limited experience — in the Caspian Sea — with offshore operations at considerable depths and even less in the difficult hydrometeorological conditions found in Arctic waters, as in the Sea of Okhotsk. Moscow's technology base is still very narrow and Soviet operating personnel often had great difficulty in making the most of sophisticated equipment acquired from the West. The Soviets would in the near future prove incapable of dealing with the pollution effects of any major blow-out or oil spillage, a problem compounded by the reportedly poor observance of safety discipline on some existing Soviet rigs.

Despite falling real prices on the world oil market and growing production difficulties in existing onshore oil-bearing regions, Soviet exports of crude oil and petroleum products to the OECD countries continued to rise in 1984, underlining the crucial role of oil as a major hard currency earner for the USSR. To some extent, Moscow attempted in the 1980s to substitute natural gas for oil exports and will continue to seek markets in Western Europe as it develops the Yamburg field in Western Siberia. The first three months of 1985 saw a 4 per cent drop in Soviet oil output over the same period in 1984 and in August 1985 the Soviets announced a 30 per cent reduction in crude oil sales to Western customers from September.[10]

In order to maintain oil output levels in the future, the USSR, in addition to applying enhanced recovery techniques onshore with related improvements in reservoir planning and engineering, has an immediate incentive to embark upon a much expanded offshore exploration effort including the Barents Sea. Almost half of the shelf area in the Kara Sea is an extension of the onshore West Siberian basin and is therefore probably highly prospective, especially for gas but possibly also for oil. The 17.5 million tons of oil produced in the Komi ASSR from the Timano–Pechora basin accounted for 3 per cent of Soviet oil output in 1980. Prospects on the offshore extension of this petroliferous basin seem more promising than onshore. The Timano–Pechora basin is bounded by the Svalbard plateau to the north and the Murmansk basin, making the Pechora and up to 800,000 km^2 of the Soviet shelf in the Barents highly prospective, at least in theory. One estimate put total recovery at up to 8 billion barrels of oil and 100 trillion cubic feet of gas.[11] Probable predominance of gas deposits and extremely harsh ice conditions in the Kara Sea render the shallow waters off the mouth of the Pechora River a high priority area for exploration, closely followed by the south-eastern and central Barents Sea.

In early 1982 Moscow announced it was considering projects in the eastern Barents in three areas of 100–300 m depth to the value of several billion dollars. The Minister of Geology referred in 1985 to a 100 per cent increase in the volume of offshore drilling in the 1986–90 five-year plan on the Soviet continental shelf at large. In 1983 the Director of the Oceanology Institute urged the formulation of a comprehensive plan for a 'scientific and industrial offensive in the higher latitudes' coupled with development of Soviet ice-resistant oil prospecting, drilling and extraction equipment to allow a rapid acceleration of exploration in the western Arctic seas of the USSR, despite the limited present knowledge of the sea-bed.

The USSR will remain reliant on Western technology inputs for work in the Barents Sea but will try to improve its own technology and equipment base over the next five to ten years. Furthermore, Moscow will require Western expertise, some degree of participation, and capital in order to offset the investment costs involved in any offshore production in the western Arctic. Soviet observers closely monitor the economic uncertainties and technical difficulties encountered by other Arctic states in their offshore programmes, especially Canada and Norway, and note the specialised equipment they have developed for their own use and to export.[12]

The Soviets first sought to gauge the interest of Norwegian shipbuilding and engineering companies in assisting offshore development of the Soviet Barents in 1982 and, following the green light from the Norwegian authorities for Norwegian firms to seek orders for equipment and expertise, contracted with Norwegian Petroleum Consultants to draw up a master plan on Moscow's behalf. Handed over in late 1984, the BOCONOR scheme included development plans and cost estimates for field exploration embracing fixed platforms and sub-sea production units, and afforded Norwegian companies some access to Soviet plans and markets in the Barents. Based on the plan, the Soviets were expected to decide what equipment and services would be provided by Soviet industry and foreign companies. Future Soviet orders could well involve more Arctic-capable jack-up and semi-sub rigs and drilling vessels for exploration and development. Most of these will probably be placed with Finnish, Norwegian and Swedish firms although a West German consortium is now trying to break into the Soviet Barents as a new opportunity to carve a place in the international offshore market.

The USSR seems especially interested in Western (UK, French and Norwegian) assistance in developing Soviet sub-sea technology. Plans are also well advanced to establish a hyperbaric centre in Leningrad, with a British firm as a strong potential partner, in order to enhance Soviet mastery of underwater repair operations. In early 1985 the Norwegian oil company Saga Petroleum was contacted by the Soviets through a 50 per cent Finnish-owned North Norwegian trading company Pomor-Nordic Trade A/S regarding technical assistance for the Barents shelf. Saga and Pomor-Nordic were encouraged to consider a joint application for an oil exploration licence in the Barents. This may be the first tentative Soviet attempt to draw foreign operators into the Soviet western Arctic offshore programme.

In 1983 Soviet officials began to resuscitate earlier vague proposals for some form of 'major Nordic project' jointly to explore and develop the Soviet Barents Sea, which would presumably furnish equipment, reduce the Soviet investment burden and assist in marketing recovered oil and gas. The Soviets floated the idea of a joint Norwegian–Soviet pipeline from the Barents, via Finland and Sweden to West European gas markets. Schemes of this nature have met gentle Finnish support. Finland has a clear incentive to offer encouragement with a view to major equipment orders and to realise its hopes for Soviet gas exports to Western Europe via pipelines across Finland and Sweden. In 1985 Finland agreed to double the capacity of its gas network, with the Soviets constructing new pipelines using Finnish equipment, as a way of solving its trade imbalance with Moscow through increased imports of Soviet gas. Rumour has it that Finnish firms were behind initial suggestions to use the Norwegian town of Kirkenes, close to the border with the USSR, as the major support base for offshore production in the Barents Sea at large.

Moscow will place great hopes on obtaining technical assistance to be financed by Western companies through low interest credits to be repaid by energy deliveries once production starts. Haggling over credit terms was what appeared to delay agreement on moving to production in the Sea of Okhotsk despite Japan's energy import needs.[13] Given current market conditions, Norwegian interest in participating in the Soviet Barents programme could well be complicated by the prospect of long-term buy-back commitments and the desire to offer credits on purely commercial terms in order to avoid future tension with Washington over East–West credit policy. Norway has already had difficulty in reaching long-term gas sales accords with Western Europe for some of its North Sea fields. Competition with the USSR for the West European gas market will increase in any case when Soviet plans to develop the Yamburg field eventually reach fruition.

Consequently, financing terms, market difficulties and sensitivity to the intra-alliance politics of technology transfer could counteract the clear interest of Norwegian firms in carving a niche in the fledgling Soviet offshore programme and may compound official concern in Oslo over the implications of such co-operation for the Soviet–Norwegian continental shelf and economic zone dispute.

The Soviet Union has so far received little interest from Oslo in long-term energy schemes for the Barents. Any Soviet hope of using co-operation in exploration to strengthen its sector claim has so far been frustrated by Oslo's determination that Norwegian equipment, expertise and presence would only be provided to the USSR for activities unambiguously clear of the disputed area. Suggestion of joint pipelines seems to indicate a Soviet tactic of seeking to move incrementally towards establishment of a *de facto* condominium regime for non-living as well as living resources in the Barents. Should major oil finds occur on the Soviet shelf, the USSR might become more flexible on the boundary question to secure a long-term Norwegian input into their development. Oslo has already signalled its readiness to consider joint exploration of structures overlapping the boundary once its location has been finally agreed. Yet formal interim arrangements permitting exploration there in the absence of a boundary agreement and establishing guidelines for later production and sharing of resources, while serving to avoid tension, will not be entertained by a Norwegian government anxious not to undermine its negotiating position.

Meanwhile, more intense prospecting by both states in the Barents in the next few years will necessitate some form of understanding on activities that affect fields extending into the disputed area. However, little progress has been achieved in shelf delimitation negotiations and Moscow is unlikely indefinitely to refrain from all probing of structures lying within or in close proximity to the area.[14] Regular bilateral consultations would provide the Soviet Union with a further opportunity to influence economic developments to the west of the sector line and perhaps occasion them to press for a minimum non-Scandinavian offshore energy presence in the Barents Sea as a whole.

The Soviets have yet to express much concern about the ecological implications of energy exploitation in the Barents. Acceptance of the conservation rationale in the Grey Zone agreement indicates some potential receptivity to Norwegian interest in bilateral co-operation to avoid damage to shared fish stocks. It is unclear whether Moscow will respond to Norway's announced intention to seek agreement on safety standards and operating procedures on oil rigs in order to avoid environmental damage in the Norwegian and Soviet economic zones.[15]

The pace of Soviet offshore exploration in the Barents and Kara Seas will mainly be determined by investment priority decisions in the 12th Five-year Plan (1986–90). The relative weight of export market conditions, availability of foreign technology and finance, and delimitation issues will no doubt be matched by very careful attention to the military implications of the ever more closely interwoven offshore tapestry appearing in the western portion of the Barents. The future pattern of political relations there is also linked to Western and especially Soviet interests in the Svalbard archipelago.

NOTES

1. Data taken from *Bulletin Statistique des Peches Maritimes*, vols. 66 and 67 (ICES, Copenhagen, 1983 and 1985).
2. Yet the agreement itself contained an explicit acknowledgement that the positions of the two sides on boundary delimitation were not prejudiced.
3. See S.V. Molodtsov, *Pravovoi rezhim morskikh vod* (Legal regime of the seas), ('Mezhdunarodnye otnoshenia', Moscow, 1982), p. 217.
4. Peter Colins, 'Swedish concern about Soviet boarding of fishing vessels in disputed Baltic waters', *Radio Liberty Research*, RL132/85 (20 April 1985).
5. Molodtsov, *Pravovoi*.
6. *Bulletin Statistique des Peches Maritimes*, vols. 66 and 67. Ministry of Agriculture, Fisheries and Food, *Statistical tables 1982* (Government Statistical Service, HMSO, London 1983), pp. 17–18, 20.
7. The Soviet formula for fisheries management around Svalbard was hinted at in P. Krymov, 'The Soviet Union and the North European countries', *International Affairs* (Moscow), no. 9 (1979), pp. 15–16. On trouble with the EC, see 'Fisheries agreement with EEC endangered', *Norinform*, no. 25 (July 1986), p. 1.
8. Sharper Soviet interest in Greenland's future political, economic and security orientation shines through an astute article by Yuri Batovrin, 'Greenland's political silence: a thing of the past', *MEMO* (Journal of the Institute of World Economy and International Relations, Moscow), no. 6, 1985, pp. 126–30. In the late 1970s Moscow unsuccessfully asked to set up a consulate in Greenland. In 1980 the Soviets offered to build a fish processing factory there to handle catches in the area from Soviet vessels. The offer was eventually declined. The potential for dispute between Iceland, Greenland (Denmark) and Norway over capelin fisheries around Jan Mayen has been reduced after a provisional agreement between the three nations in Autumn 1986 (*News from Iceland*, November 1986, p. 12).
9. David Wilson, 'Latest developments in the Soviet offshore programme', *Petroleum Review*, April 1985, pp. 11–12.
10. The *Guardian*, 20 August 1985. See also Philip Hanson, 'Soviet oil deliveries: charity begins at home', *Radio Liberty Research*, RL293/85 (6 September 1985).
11. A Meyerhoff, 'Petroleum basins of the Soviet Arctic', *Geological Magazine*, vol. 117, no. 2 (1980), pp. 101–86.
12. See, for example, M.P. Krasnov, 'Exploitation of the Canadian shelf', *S.Sh.A.* (Journal of the Institute for the Study of USA and Canada, Moscow), no. 2, (1981), pp. 87–94; O. Kazakova, 'The Scandinavian concept of an "industrial niche"', *MEMO*, no. 6, 1984, pp. 53–65; Yu. Denisov, 'The international significance of Soviet–Finnish Cooperation', *MEMO*, no. 2 (1985), pp. 21–31.
13. *Argumenty i fakty*, no. 9 (28 February 1984), p. 6; *Petroleum Economist*, October 1984, p. 387.
14. Finn Sollie, 'Oil and foreign policy', *Nyhetsbrev*, no. 4 (1984), p. 20.
15. *Izvestia*, 6 March 1985 gave details of a new Soviet decree on environmental protection in the Soviet far north and northern seaboard coastline. The prospects appear good for a Soviet–Norwegian environmental

protection agreement embracing the frontier area between the two countries, but mainly applying to transboundary air pollution, *Norinform*, 27 August 1985, p. 9.

6

Soviet Military Strategy and Northern Waters

Tomas Ries

Soviet military strategy in Northern Waters can on the whole be said to consist of two main related parts. First are those primary Soviet strategic military interests which make the region of importance to the USSR, and the associated strategic forces which are allocated to the region to achieve these primary objectives. These forces have an extra-regional orientation and on the whole only affect the regional military situation indirectly, by raising or lowering the general strategic interest of the superpowers in the area. Secondly, however, these strategic interests create a series of secondary regional theatre-level military support requirements. These consist of those operations, and their associated forces, which are necessary to protect the strategic forces and safeguard their operational viability in wartime. These support requirements have a strongly regional orientation and decisively affect the local military equilibrium.

Both are important. The first determines the level of superpower strategic interest in the area, and thus the scale of the military involvement. By examining these interests we can understand why the USSR is active in the area, which in turn allows us to assess the impact of current strategic developments on the region and assess the prospects for the future. The second set of support requirements determines the actual objectives and scope of the regional theatre-level military operations which affect the local military situation. Examining these can give an idea of the existing level of military threat and of the evolution in the regional equilibrium of military power.

SOVIET NORTH-WESTERN MILITARY STRATEGIC INTERESTS

The north-western USSR and adjacent areas are of military strategic importance for the USSR for five main reasons. These are, in presumed order of importance:

1. The Kola and White Sea coasts are — so far — the best basing areas for the Soviet strategic submarine (SSBN) forces, and the adjacent Arctic waters constitute the optimal operational concealment and launch stations for these forces. Thus, a majority of Soviet SSBNs — 60 per cent of the total force — are based here. Of these, all of the most modern vessels, built since 1971 — *Delta* I, II and III and the *Typhoon* class — have the Arctic waters in the Barents Sea, Kara Sea and Arctic Ocean itself as their operational concealment and launch areas.

2. The Arkhangelsk Air Defence Sector (ADS) is vital for Soviet strategic air defence, lying beneath the shortest transit route between the Soviet industrial, demographic and military heartland, and the continental US strategic bomber and ICBM bases. Secondly, the Kola area itself, because of its SSBN bases, constitutes a major strategic counterforce target which calls for air defence. Thus, the USSR has allocated an important part of her strategic air defence forces to this sector and has given it priority for the most modern weapons systems in this field.

3. The Norwegian Sea forms an important launch area for Western seaborne nuclear attack forces (carrier-borne aviation in the 1950s and SLBMs since then) making this an important area for Soviet defensive strategic anti-submarine warfare (ASW). While the importance of this mission and the forces allocated to it have declined considerably since the change in Soviet naval strategy in favour of defence of Arctic SSBN Ocean Bastions in 1971 and since the deployment of the long-range *Trident* in 1980, the threat posed by the remaining Western C-3 *Poseidon* SLBMs with launch stations north of the Greenland–Iceland–Faroes–United Kingdom (GIFUK) Gap remains sufficiently large to continue making this an important task for the Northern Fleet. In the coming years the priority of this mission may increase once again to counter the increasing deployment of the USN *Tomahawk* SLCM.

4. The north-western Arctic coastline constitutes a vital forward operating and refuelling area for Soviet long-range strategic bombers *en route* to continental North American targets. Considering the relatively small Soviet strategic bomber forces and their aged present systems, this is probably a relatively minor consideration in Soviet strategic planning. However, this may change with the imminent introduction of three new weapons systems in this service and possible

Soviet reactions to the Strategic Defense Initiative (SDI), which could involve an increased emphasis on strategic bombers.

5. The Kola offers the best basing facilities providing access to the central Atlantic for the Soviet naval forces which are assigned the task of cutting NATO Atlantic Sea lines of communication (SLOCs). This may not yet constitute a major Soviet strategic objective. As the Soviet Navy continues to grow, its operational presence will expand further, and will begin to extend into the central Atlantic. At this stage the maintenance of a powerful conventional sea control presence there, with all the attendant psychological pressures on Western Europe which this entails, may come to constitute a major Soviet strategic objective in itself.

Each of the above points will now be examined in greater detail.

The Kola and Arctic in Soviet SSBN strategy

Soviet SSBN basing and operations requirements probably constitute the single most important factor for the strategic value of the Kola and adjacent Arctic waters, for two reasons. Firstly, the SSBN forces have a high value in Soviet nuclear strategy. This is indicated both by the relative size of the SLBM force in the Soviet strategic arsenal, comprising 40 per cent of strategic nuclear delivery vehicles (see Figure 6.1) and by the high allocation of research and construction resources towards the SSBN force. In the last ten years, four new SSBNs (the *Delta* II in 1975, the *Delta* III in 1976, the *Typhoon* in 1981 and the *Delta* IV in 1984) have been launched, and five new SLBMs (the SS-N-17 in 1977, the SS-N-18 models 1, 2 and 3 in the late 1970s and the SS-N-20 in 1981). This is the only strategic nuclear force which is relatively secure from a surprise first strike or from attack during a nuclear war. Thus, it represents a strategic reserve which can be held back either for actual use or for use as a bargaining chip in a post-exchange phase. In the coming years one further factor adds to the importance of the SSBN force in Soviet strategy. In the event that the USSR perceives a real threat to the credibility of her ICBM forces from the SDI she may decide, as a partial remedy, to increase the number of her other nuclear delivery systems which are not affected by the space defences.

One such force would be the short-range SLBMs. If launched close to their US targets and with a depressed ballistic trajectory, these could bypass a majority of the space defences and increase the chances of penetrating to their targets. They would also substantially reduce

Figure 6.1: The evolution of Soviet strategic delivery vehicles, 1960–84

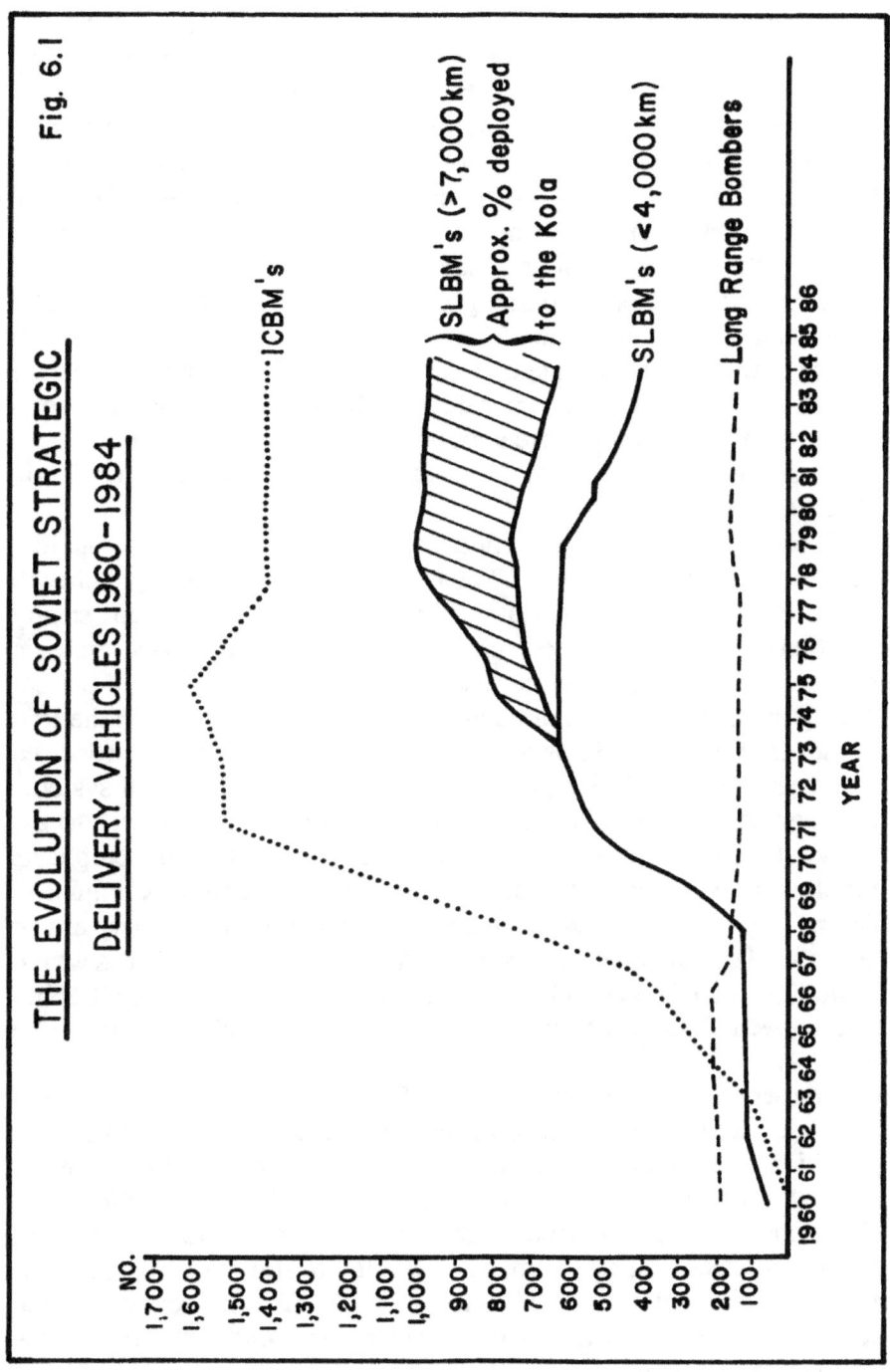

warning time. Should this be one of the consequences of the SDI, then the value of the SSBN force for the USSR would increase even further.

Secondly, the SSBN forces are specifically important to the Kola and Arctic because these represent prime Soviet basing and operating options for these forces. Today about 60 per cent of Soviet SSBNs are based on the Kola (with the remaining 40 per cent on the Kamchatka Peninsula). Of these forces, all of the most modern systems (*Delta* I, II, III and *Typhoon*) or roughly 60 per cent of the Kola SSBN ORBAT (order of battle), have their operational launch stations in the Arctic waters off the Kola, and it has been reported that, since 1975, none of these has been detected passing south of the Norwegian Sea. The remaining 40 per cent of the SSBNs consist of *Yankee* class vessels, whose limited 2,400 km range of the SS-N-6 SLBM requires transit to launch-stations in the western Atlantic. However, it is reported that a majority of these systems are kept in reserve in the northern area, and only a smaller force of two to three vessels are on station in the Atlantic at any one time.

Figure 6.1 shows the general evolution of Soviet strategic nuclear delivery systems between 1960 and 1984, indicating the rapid growth of the SSBN forces in the later 1960s, and their subsequent steady position as the second most important force in the Soviet strategic nuclear arsenal. The development of the Soviet sea-based component during these years can be divided into four general phases. In the first phase, from the late 1950s to 1960, neither the priority accorded the submarine-based missiles, nor the capabilities of the actual systems themselves, made this service of primary strategic importance to the Soviet Union. At this stage the overall Soviet strategic emphasis involved drastic cuts in the navy, and its primary role was limited to defence against the threat from carrier-borne Western nuclear attack aircraft. Those submarine-based nuclear ballistic missile forces which it did have consisted of a limited force of 17 *Zulu* V and *Golf* I SSBs and *Hotel* I SSBN, all armed with the very limited 240 km range SS-N-4.

The second phase from 1960 to 1967 is characterised by a revision of Soviet strategy which placed a high strategic value on the force, but in which the capabilities of the systems in use remained extremely limited. The new policy was announced in January 1960, when Khruschev stated that henceforth the Strategic Rocket Troops and nuclear missile-bearing submarines would constitute the main arm of the Soviet armed forces. This led to a significant increase in the resources allocated towards the development and building of new SSBNs and SLBMs.

Figure 6.2: Soviet SSBN basing and launch areas, 1959–68

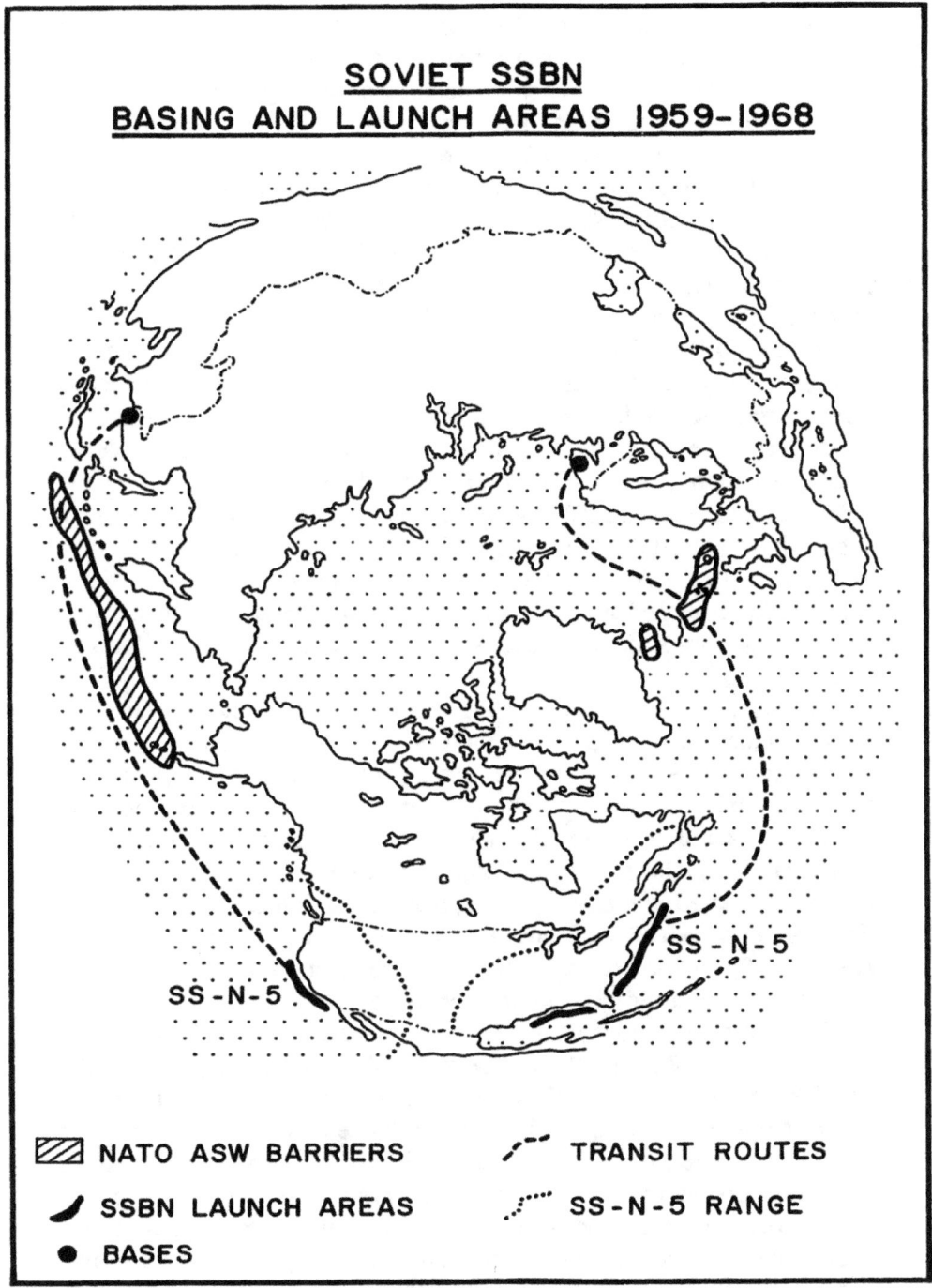

The third phase, between 1967 and 1972, is characterised by the entry into service of the first of these new systems, which significantly improved the credibility of the SSBN force, but where the range limitations of the new missile still obliged the SSBNs to transit through Western ASW barriers to reach launch areas (see Figure 6.2). The first of the second generation of Soviet SSBN/SLBM systems, the *Yankee*/SS-N-6 combination, became operational in 1967. This system was considerably improved in terms of reliability and capabilities. The 2,400 km range of the SS-N-6 meant that for the first time the SSBN force could cover all strategic targets in the US. However, it did require transit to launch areas in the western Atlantic and eastern Pacific, which made it vulnerable to US ASW efforts along strategic choke points (the GIFUK Gap and the Aleutians) and therefore of uncertain strategic utility. On the other hand it also represented the first secure Soviet strategic nuclear reserve, which, if held back in protected waters, could be used as a bargaining chip or for use in a post-exchange phase. As such it must have ranked highly in Soviet planning, as indicated by the considerable growth of the SSB/SSBN force from 37 submarines with 112 SLBMs in 1967 to 62 submarines with 576 SLBMs in 1972. At this stage, primarily the Kola, and secondly Kamchatka, emerged as the major SSBN basing areas, mainly because of their relatively unimpeded access to the launch areas off the US coast (see Figure 6.3).

The fourth phase started in 1972 and continues to the present day. This is characterised by two factors: firstly by a growth in the proportion of SLBMs in the Soviet armoury and a probable corresponding increase in the importance of this force; and secondly, by an increasing operational orientation to the Arctic waters in the vicinity of the Kola (see Figure 6.4). This has had the consequences of increasing the Kola's basing value, and of making the Arctic waters themselves to be of vital strategic importance. This has drawn in American SSNs for strategic ASW purposes, which in turn could lead to an escalation of Soviet forces in and over the Arctic.

The first of the new systems became operational in 1973 when the *Delta* I/II/III, and, equally importantly, the various models of the SS-N-8 and SS-N-18 SLBMs, were deployed. The intercontinental range (between 6,500 and 9,100 km) of these missiles meant that for the first time the Soviet SSBN forces did not have to transit Western ASW barriers but could deploy to launch stations in the vicinity of the Soviet coasts, where, in addition, Soviet naval and air forces could help protect them from Western surface and air strategic ASW activities. During this phase the SSBN force continued to grow rapidly, reaching

Figure 6.3: Soviet SSBN basing and launch areas, 1968–73

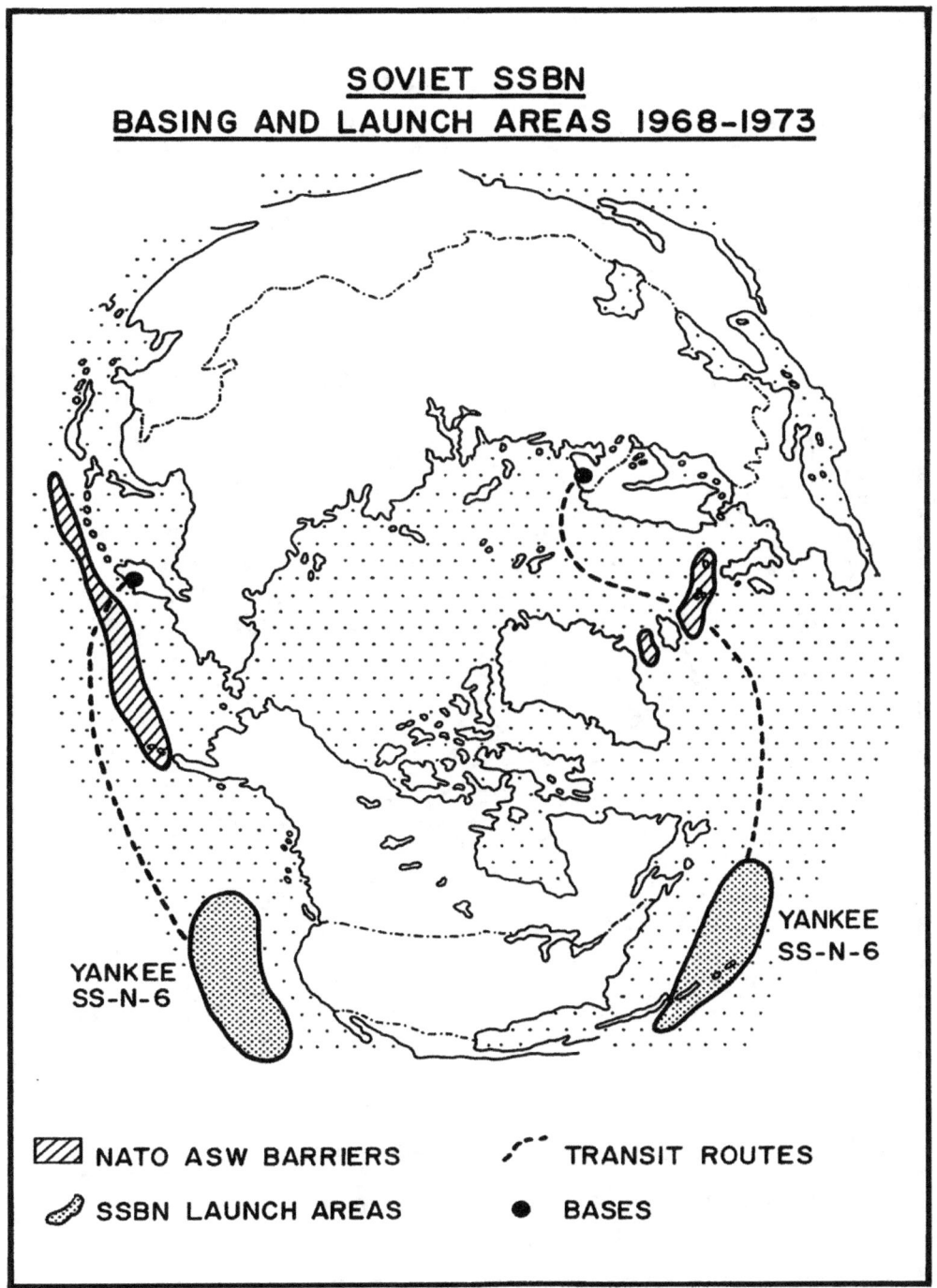

Figure 6.4: Soviet SSBN basing and major launch areas, 1973–85

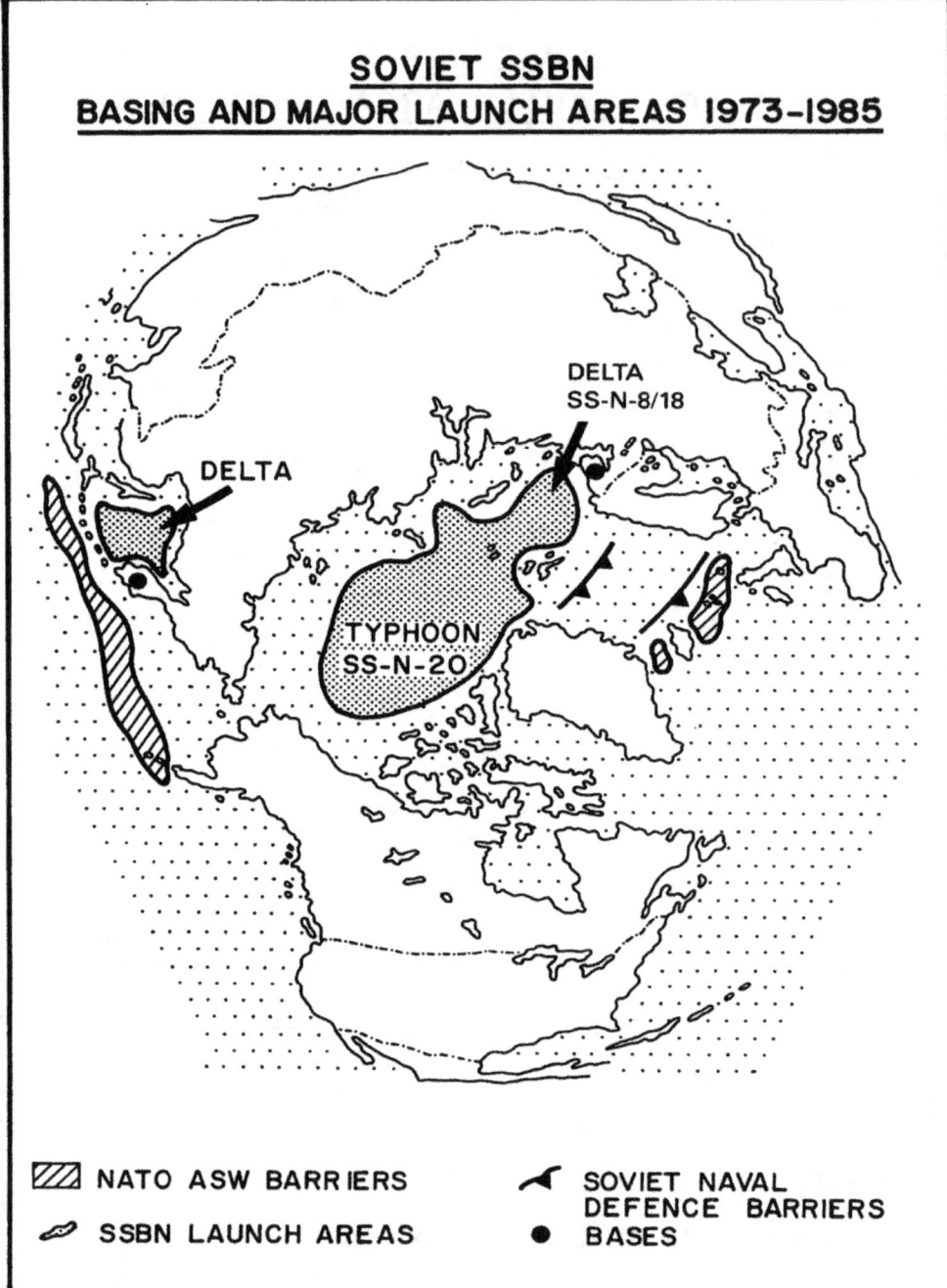

82 SSB/SSBNs equipped with 991 SLBMs by 1981. The introduction in 1981 of the *Typhoon* SSBN, the first Soviet SSBN specifically designed for Arctic under-ice operations, seems to confirm the trend towards Arctic operations, which has made the Kola an important basing area, with, since 1977, roughly 60 per cent of the Soviet SSBNs, and, since 1981, all of the most modern *Typhoon* class SSBNs.

Strategic air defence

This is perhaps the second most important Soviet military strategic interest in the north-west at present, and consists of two main efforts: air defence against the long-range bomber threat, and Early Warning (EW) and defence against ICBM attack. Neither of these efforts probably ranks among the top Soviet strategic concerns at the moment, but both are sufficiently important nonetheless to affect the strategic importance of the far north.

The first of these missions is maintained for two reasons: because of the relatively low but consistent level of threat presented by the US strategic nuclear bombing forces, and because of the location of the Kola and Arkhangelsk coastal areas beneath the direct transit routes from continental US bomber bases to the Soviet countervalue and counterforce heartlands.

The overall degree of threat from the US long-range strategic bomber force has remained fairly consistent over the last 25 years, with 12 per cent of the total US strategic nuclear delivery force in 1984 consisting of long-range bombers, compared with 12 per cent in 1959. This element of the nuclear threat, although not one of the major threats, has nonetheless made strategic air defence an important part of Soviet nuclear strategy. The importance attached to this is further borne out by Soviet attitudes to the air defence forces. They rank as the third highest branch of the armed forces, and have — with one exception in the 1970s — consistently received a high priority for receiving new and modern equipment. Between 1950 and 1970, 15 of the 21 tactical aircraft deployed in the Soviet air forces were designed specifically for air defence, and in the five years to 1986 the air defence forces have received three new interceptor aircraft (MiG-25E in 1980, MiG-31 in 1983, MiG-29 in 1984) and currently two new aircraft, the Il-76 AWACS and the Su-27 interceptor, are expected to join its ranks. This service is one of the most favoured among the air forces for the reception of new type designed aircraft.

Secondly, a large proportion of these aircraft is assigned to the north-west, to the ADS (which appears to correspond roughly to the Leningrad Military District in terms of the area it encompasses), and this area appears to have priority for the reception of the most modern aircraft. This is partly because of the perceived importance of this region as a strategic bomber transit route. Figures 6.5 to 6.8 show the general evolution of the US ICBM and strategic bomber forces in 1960 and 1985 and their flight paths in relation to major Soviet countervalue and counterforce targets respectively. In each of the maps the thickness of the arrows indicating the flight paths of the respective systems shows the relative proportion of forces which are estimated as following that particular flight path. Figure 6.5 shows the principal flight paths of the US long- and medium-range strategic bomber forces in 1960 between their principal bases and the main Soviet counterforce targets, while Figure 6.6 shows the same situation, with the addition of US ICBMs, in 1985. Figures 6.7 and 6.8 show the same but in relation to Soviet countervalue targets. While there is a certain difficulty in making estimations of this kind, due to the classified nature of the subject, and while one must allow for variations in launch scenarios, it appears that in both cases a significant proportion of both long-range bombers and ICBMs would transit over the vicinity of the Soviet north-west. Soviet targets have to some extent shifted east of the Urals, and US long-range bomber bases have been spread around the globe to such areas as Diego Garcia in the Indian Ocean, Guam and the Philippines, but on the whole the majority of targets remain west of the Urals, and the majority of launch areas and bomber bases in the continental US. This continues to make the north-western USSR a major transit route for these systems and, as such, one of the most forward placed areas for the Soviet strategic air defence effort.

In the years to come this importance will probably grow. Both the imminent deployment of the US B–1 strategic bomber, and the deployment in 1982 of the AGM-86B Air-launched Cruise Missile (ALCM), indicate the continuing and, from the Soviet perspective, growing threat from the manned bomber systems. In addition, the 2,500 km ALCM is increasing the need for Soviet long-range Arctic air interception, as the destruction of the bomber prior to ALCM launch is one of the few means presently available to the USSR to counter this threat. This means that interception must be carried out beyond the 2,400 km range of the ALCM, which enhances the need for Arctic forward interceptor and radar bases.

However, a second reason for the priority of the Arkhangelsk ADS may be because of the other missions which the air defence has in this area. Thus, it is also responsible to a certain extent for providing air

Figure 6.5: Major flight paths of US airborne strategic systems to Soviet counterforce targets, 1960

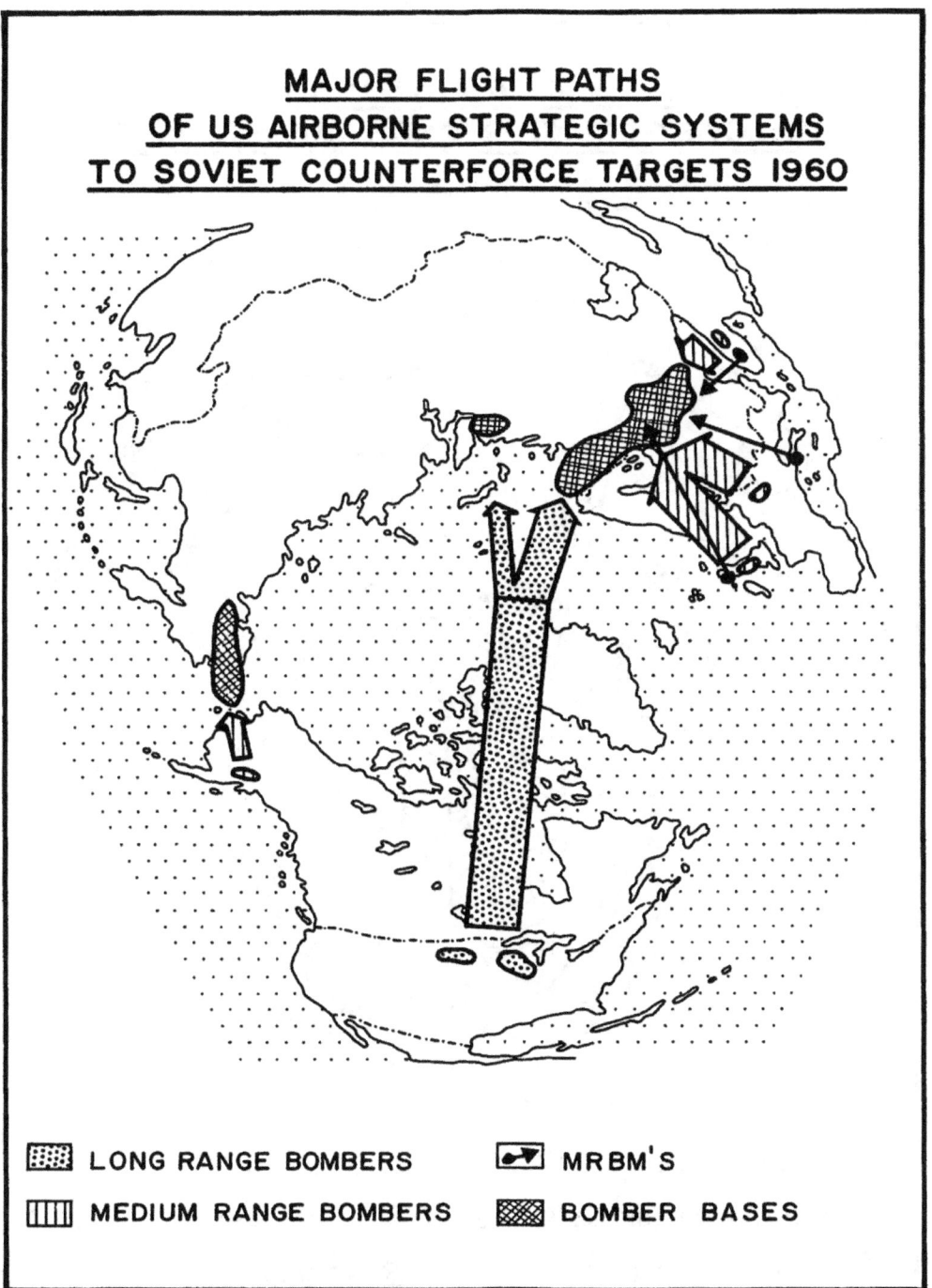

Figure 6.6: Strategic airborne transit routes to Soviet counterforce targets, 1985

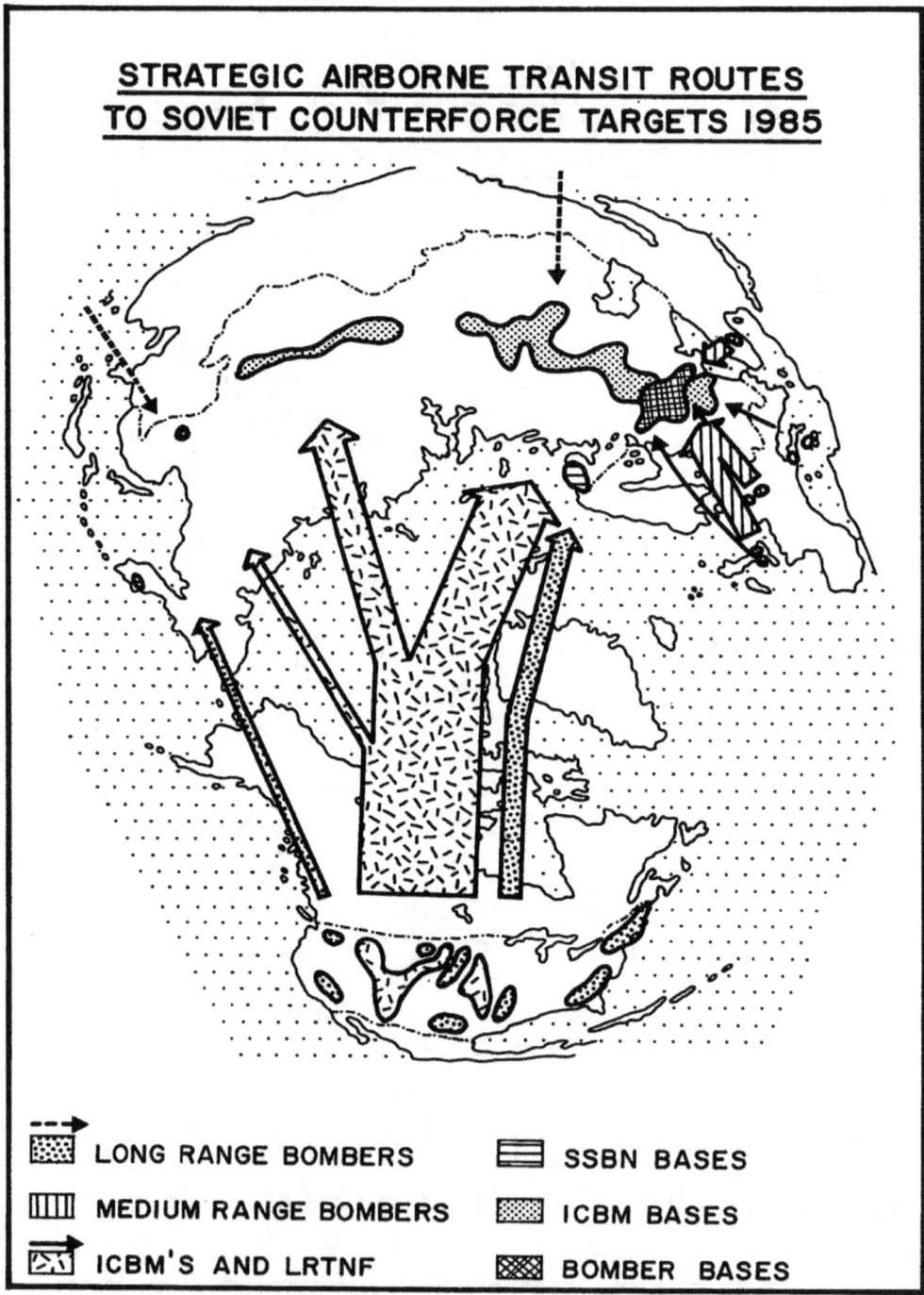

Figure 6.7: Strategic airborne transit routes to Soviet countervalue targets, 1960

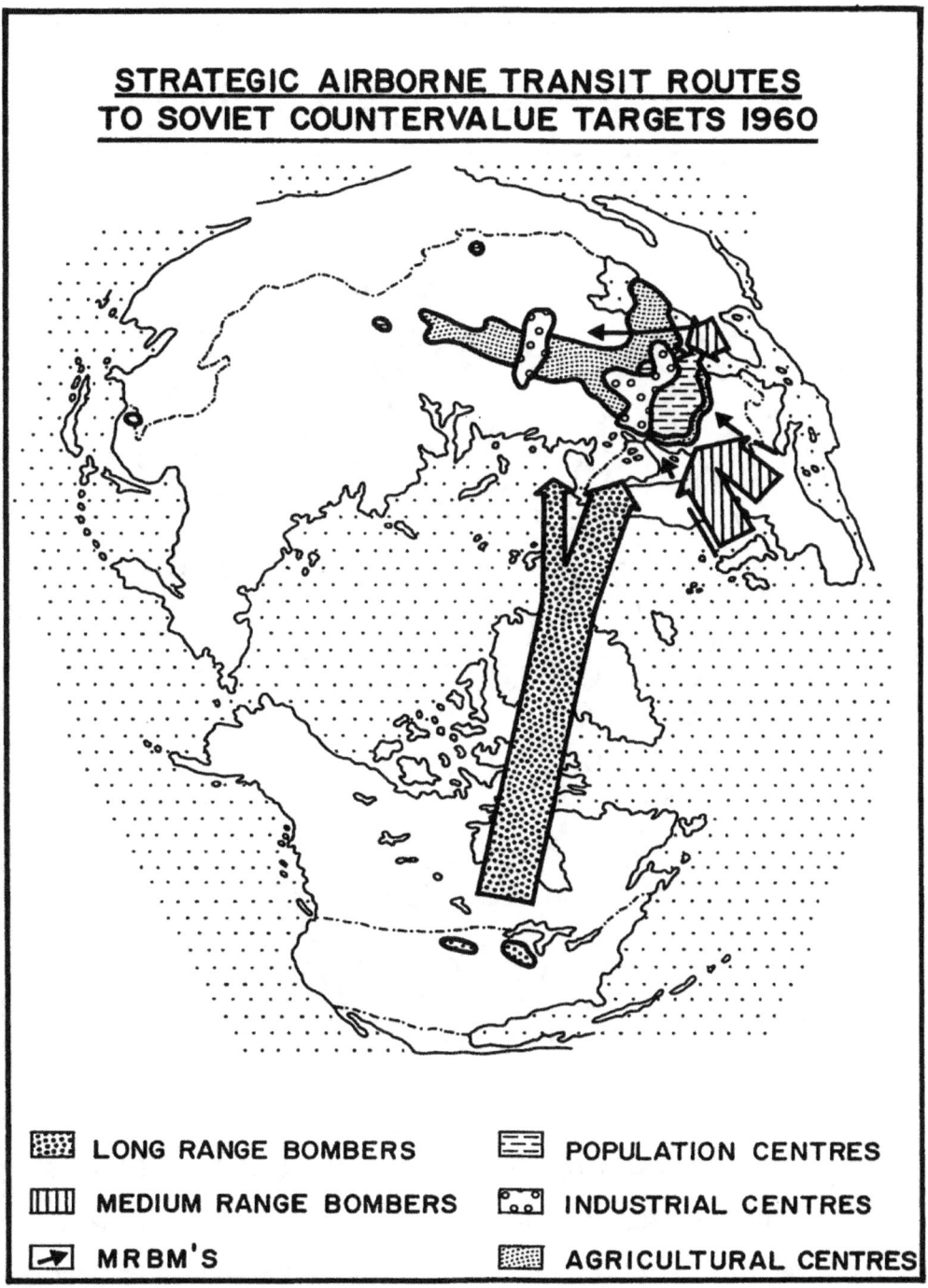

Figure 6.8: Strategic airborne transit routes to Soviet countervalue targets, 1985

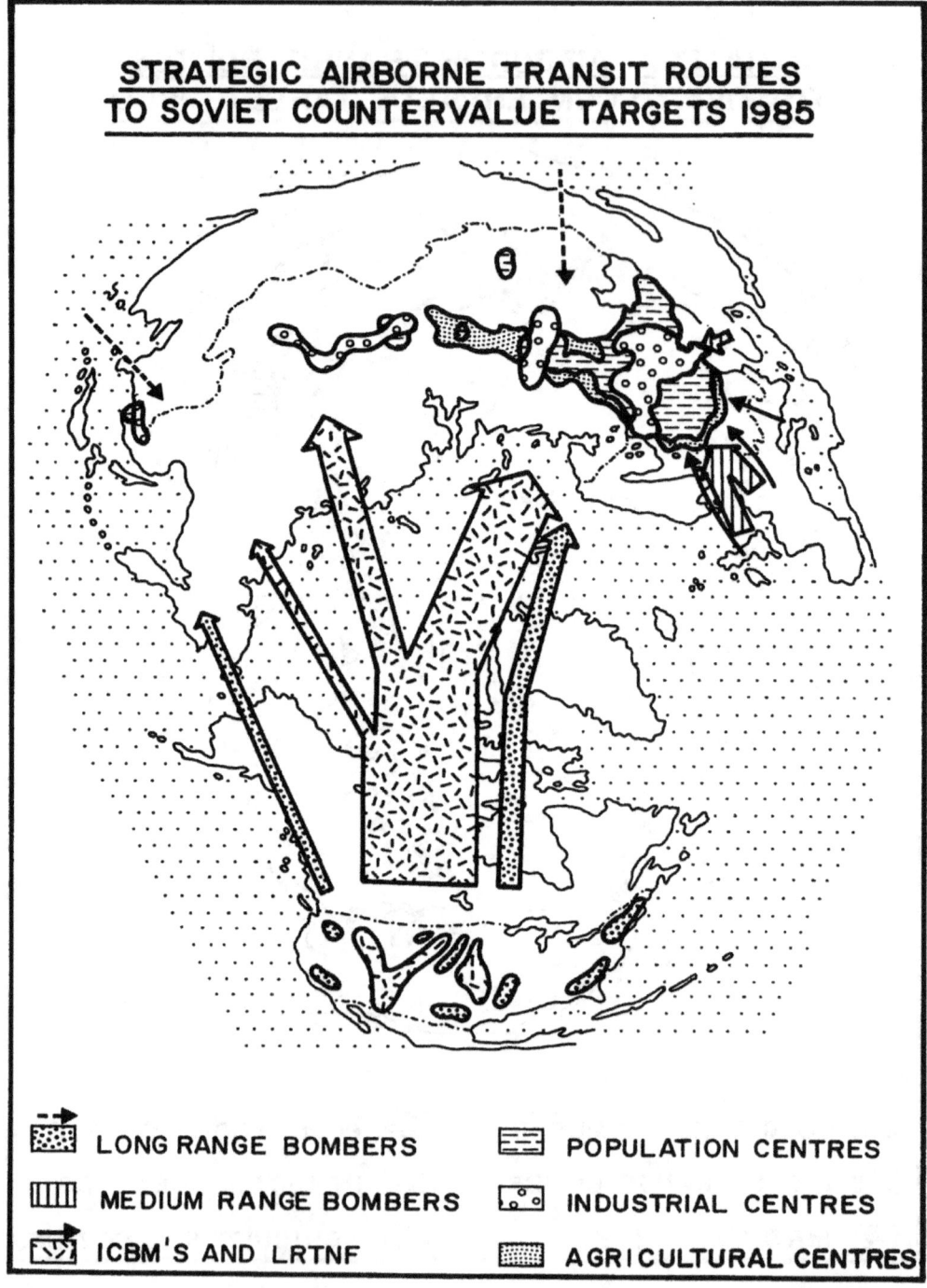

cover for the Northern Fleet in its task of defending the Arctic SSBN Ocean Bastions, and, since 1981, for air cover for the ground forces in the North-western Theatre of Military Operations.

The second set of strategic air defence considerations conferring strategic value to this area is the Soviet Ballistic Missile Early Warning (BMEW) and Ballistic Missile Defence (BMD) efforts. The first of these two is no doubt a high Soviet strategic priority, and the north-western Arctic coastline from the Kola to the Yamal Peninsula forms an important basing area for the long-range BMEW radars as it is the closest area facing the incoming flight paths of the ICBMs. As a result a number of the BMEW radars have been placed here. It appears that there are two of the older *Hen House* radars in this region, and that two of the modern phased array radars have also been placed here. The area would also be important to Soviet BMD efforts as a forward location for early warning and target tracking radars. This is a secondary function of the *Hen House* systems, and possibly of the new phased array radars. However, the Soviet attitudes towards BMD are not so clear. It would appear that the principle of strategic defence rates highly in Soviet planning (and probably even more so now that the US has launched its SDI programme), but the means for carrying it out, in the form of a viable ABM and radar tracking system, have not yet been established. Thus, this strategic factor is presently probably a relatively minor influence on the northern strategic environment, but it is an area in which considerable changes could occur if the Soviets manage to develop effective defences against ballistic missiles.

Strategic ASW north of the GIFUK Gap against the SLBM and SLCM threat

The strategic importance of this mission has probably declined considerably in the last ten years, following the change in Soviet naval strategy in the early 1970s towards defence of the Soviet Arctic SSBN Ocean Bastions (see p. 96 above) and away from the increasingly daunting task of strategic ASW. This second mission had previously been the major task of the conventional forces of the Soviet Navy, but had constantly eluded the Soviets as the ranges of Western sea-based nuclear delivery systems grew. By the early 1970s, when it became known that the US was planning to deploy the 7,400 km range *Trident* SLBM, the increasing impossibility of this task probably convinced Soviet planners to abandon it as a strategic priority. At the same time the first Soviet long-range SSBNs appeared, and the forces

Figure 6.9: The evolution of Soviet northern naval operations for defence against the seaborne nuclear strategic threat

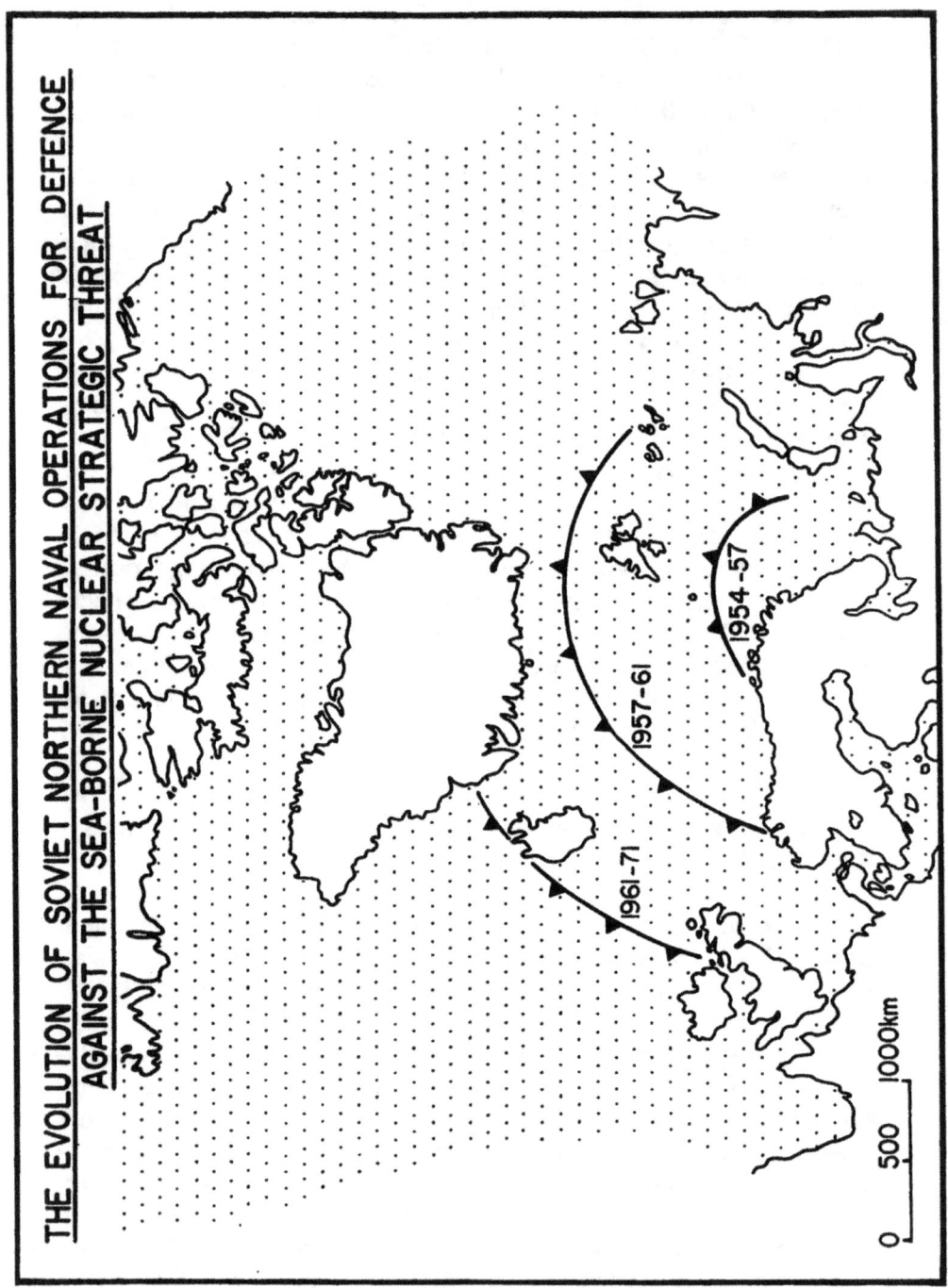

assigned to strategic ASW were probably reassigned for SSBN protection. However, this may still change slightly. With the introduction of the 2,400 km US Navy SLCM, the Norwegian Sea again becomes an important launch area, and if the Soviets perceive the SLCM as a potential strategic threat — which they probably do — they may again increase the allocation of forces for defence against this threat.

However, as the roughly 15-year (1954–71) Soviet quest for strategic defence in the Northern Waters is primarily responsible for the present large conventional order of battle of the Northern Fleet, and as most of the forces and operational modes developed for strategic ASW have been carried over to the new primary tasks of defending the Arctic SSBN Bastions, an outline of the evolution of the strategic naval defence efforts is useful (see also Figure 6.9).

Initially, in the mid-1950s, the defence against the seaborne strategic nuclear threat constituted the primary *raison d'être* of the Soviet fleet. The threat then consisted of carrier-borne aviation with fairly limited 900 km range limits. Soviet defence efforts consisted of trying to deny the launch areas to the Western forces through co-ordinated attacks by air, surface and diesel submarine forces operating within the range of land-based air cover. During this phase, the Soviet forward line of defence lay roughly along the edge of the Barents Sea.

In 1957 the US Navy deployed the A-3D long-range (c. 2,000 km) attack aircraft. This placed the carrier-borne threat beyond the range of land-based air cover, making the use of surface forces ineffective and forcing a shift in Soviet naval strategy towards the use of nuclear-powered attack submarines and naval aviation anti-shipping aircraft for sea-denial operations in the new threat areas. This resulted in major cuts in Soviet naval construction for most other types of forces. The new Soviet forward barrier lay roughly along the northern Norwegian Sea, and Soviet exercises indicated that they believed the main battle would be waged off the Lofoten Islands.

By 1959 the threat had expanded once more with the deployment of the first *Polaris* A-1 SLBM. The increased 2,700 km range of the missile (reaching most major Soviet strategic targets west of the Urals), combined with the impossibility of intercepting it once launched, made this a major strategic concern. At the same time, the longer range and submerged launch mode required new methods of defence against the seaborne threat. This led to two changes in Soviet naval strategy: increasing the range of the outer line of defence, and introducing a return to an attempt to build up a balanced fleet.

The previous relatively simple sea-denial operations were no longer sufficient. Defence against the SLBM threat required strategic ASW, which implied limited control of the sea areas where this was to be carried out in order to permit the Soviet surface and air ASW forces to operate. This appears to have led to a decision to build up a more balanced navy, capable of providing the sea control forces necessary for protecting the main surface and air forces involved in strategic ASW. Operationally, it led to a further expansion of Soviet naval defences in the Norwegian Sea, with an attempt to establish an Outer Zone of Defence stretching down to the southern Norwegian Sea, seeking to deny this area to the Western forces, and an Inner Zone of Defence along the edge of the Barents where absolute superiority was sought.

In 1964, with the deployment of the first *Polaris* A-3 4,600 km-range SLBM, this effort was reinforced with an apparent decision to continue to push forward the Outer Zone of Defence to this 4,600 km limit, while expanding the Inner Zone to the previous Outer Zone boundaries. This led to the allocation of resources for further large surface sea-control vessels designed for forward operations. The Outer Zone of Defence now lay roughly along the GIFUK Gap, and the Inner Zone edge along the northern and central Norwegian Sea.

However, by 1971 the task of strategic ASW was probably relegated to a lower priority for two reasons. In the first place it became known at that time that the US was planning to deploy the 7,400 km-range *Trident* SLBM. This could operate in such vast expanses of the ocean so far from the USSR that strategic ASW against it was probably impossible. Secondly, the USSR had deployed her own first intercontinental range SLBMs, the SS-N-8, and for the first time could try to preserve her SSBNs in safe waters close to the USSR. It was therefore probably decided to shift the main conventional priority of the Navy towards defence of the SSBN Ocean Bastions, and away from strategic ASW. It seems that this forms the current priority task of the Northern Fleet which, should the US implement its Arctic under-ice ASW programme, could grow in importance in the future.

The Arctic coastline and the Soviet strategic long-range bomber force

For the last 25 years the long-range strategic bomber force has been of relatively low importance in Soviet strategic planning, and so this factor adds relatively little to the strategic importance of the north. This is borne out both by the small size of the force, and the — until recently — relatively scarce allocation of resources to this service, leaving it with ageing aircraft. From a peak strength in 1960, when the long-range bomber force constituted 68 per cent of the Soviet strategic nuclear force, it rapidly dropped to 33 per cent by 1967, and by 1971 reached 7 per cent, at which point it has remained to date. During this period its basic aircraft, the *Bear* and the *Bison* — both dating from 1956, though they have been modernised — remained the same, indicating the relatively low priority this service has had until about ten years ago.

This is now changing. While it is debatable to what extent the *Backfire* is intended for use as part of the long-range intercontinental nuclear forces, a number of other new systems are on the verge of deployment. These include the substantially updated new *Bear* H intercontinental bomber, the entirely new *Blackjack* strategic intercontinental bomber and the AS-15 ALCM, all of which are soon to enter service. This indicates that the Soviet long-range bomber force will probably at the least remain at its present level in the years to come, and it could also increase. One possibility is for the Soviets to counter a perceived US strategic advantage derived from an active SDI system by increasing the proportion of their nuclear weapons delivered by intercontinental bombers. Especially if the trials with the AS-15 system are successful, this could lead to a significant future increase in the bomber forces.

Should it occur, the strategic value of the Arctic coastline will increase, as it represents a forward refuelling and support basing location for the strategic bombers. A number of these airfields are located in the north-west, on the Kola and on the Arctic islands of Novaya Zemlya and Zemlya Frantsa Iosifa.

The Kola as a basing area for Soviet conventional naval forces assigned the task of cutting NATO Atlantic SLOC

This factor as a major Soviet strategic interest is a tentative suggestion, as it is not yet certain to what extent it constitutes a major Soviet strategic interest on a par with the others listed here. For the present, it would appear to be secondary to other major Soviet conventional naval missions, such as the primary task of protecting Soviet SSBNs in their Arctic launch areas and the second ranking mission of carrying out strategic ASW north of the GIFUK Gap. However, Atlantic SLOC interdiction could form a major Soviet conventional naval objective in the years to come. It is worth noting that the share of naval resources allocated to the anti-shipping role appears to have grown considerably recently. Among attack submarines, six new types are relevant since 1980: the *Kilo* class SS (1980); the *Oscar* class SSGN (1981); the *Mike* class SSN (1983); the *Sierra* class SSN (1984); the *Yankee* class SSN; and the *Akula* class SSN (1985). In addition, roughly 50 per cent of all *Backfire* B have been allocated to the Naval Aviation in the anti-shipping role, further indicating the importance of this mission. It may be that a major part of these forces is intended to support the defence of the SSBNs, and would operate in a defensive sea-denial role in the Norwegian Sea and GIFUK Gap, trying to keep Western strategic ASW forces at bay. However, the advantages of cutting western Atlantic SLOC make this an attractive alternative possibility.

In this context there are two types of operation which could be of value to the USSR. In the first case, there is the more classic wartime operation of cutting Western Europe off from reinforcements from North America, which, in the event of a protracted non-nuclear war, could be critical. This would primarily involve the type of sea-denial forces listed above. Secondly — and more importantly in the long run if war does not break out — is the effect which a permanent Soviet naval presence among the central Atlantic shipping lanes would have in peacetime. This would call for the development of sea control forces, such as those emerging with the development of the *Kara* class CGN and the construction of an aircraft carrier in the Black Sea. Should the Soviet Union be able, in the long term, to build up a sufficient naval force and expertise to enable her to establish a permanent Atlantic sea control force, it could have important psychological effects on Western Europe, where there might be the perception that it was being cut off from the US. This may prove an important incentive for developing such a force, which could be based mainly with the Northern Fleet.

SOVIET STRATEGIC MILITARY SUPPORT REQUIREMENTS IN THE NORTH-WEST

The previous section outlined the main reasons why the Soviet north-west and adjacent areas are of strategic interest to the USSR.

These are summarised below, together with the main forces responsible for their prosecution.

Figure 6.10: Soviet military perception of the Northern Waters area

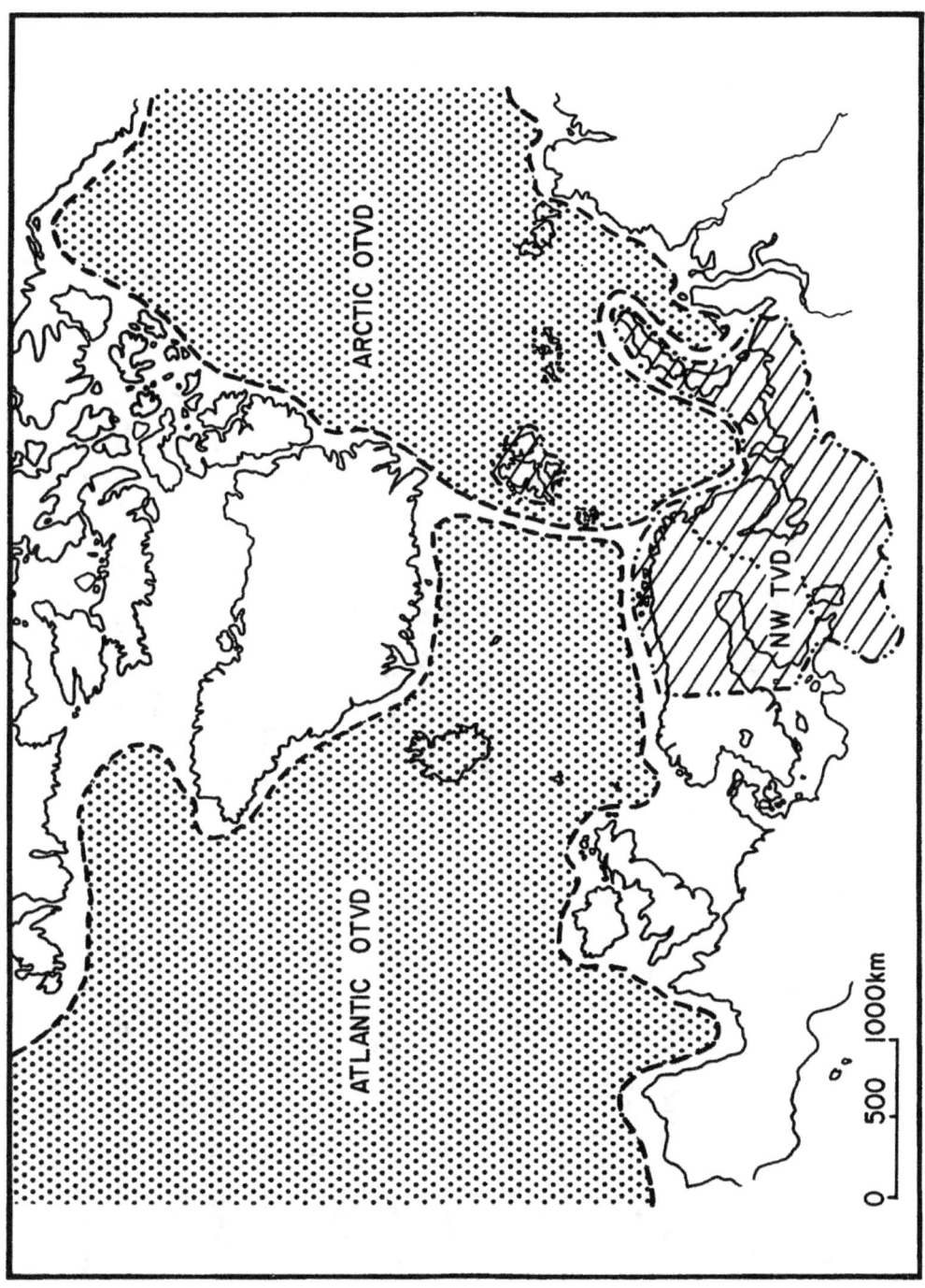

Table 6.1: The translation from major strategic objective to strategic operational task

Strategic objective	Operational task	Service with primary responsibility
1. *Strategic nuclear offensive*		
1.1 With long-range bombers:	Penetrate to target areas:	VVS: DA
1.2 With SSBNs:	Secure deployment to, and standby in, launch areas	Navy
2. *Strategic air defence*		
2.1 Against long-range bombers/ALCM:	Early warning of attack:	Vojska PVO
	Prevent bombers reaching targets:	Vojska PVO
2.2 Against ICBMs:	Early warning of attack:	Vojska PVO
	(Prevent missiles reaching targets?:	Vojska PVO)
2.3 Against medium-range bombers:	As for long-range bombers	Vojska PVO
3. *Strategic sea defence*		
3.1 Against SSBNs:	Prevent SSBNs from launching SLBMs:	Navy
3.2 Against SSGNs:	As for SSBNs:	Navy
4. *Strategic naval conventional*		
4.1 Isolate W. Europe from US:	Establish CENTLANT presence: (peacetime)	Navy
	Cut Atlantic SLOC: (wartime)	Navy
4.2 Increase influence in Third World:	Maintain fleet in Third World: (peacetime)	Navy

Table 6.2: The translation from strategic operational task to regional strategic support requirements

Strategic Operational task	Regional northern support requirement	Service with primary responsibility
1. Strategic nuclear offensive		
1.1 LRB penetration to target areas:	Protect Arctic forward airfields:	Vojska PVO
	Provide Arctic long-range escort:	Vojska PVO
1.2 Secure SSBN deployment to, and standby in, launch areas:	Protect Northern Fleet Arctic SSBN bases:	Navy and Vojska PVO
	Protect LR SLBM Arctic launch areas:	Navy
	Assist MR SLBM SSBNs reaching WESTLANT:	Navy
2. Strategic air defence		
2.1 LRB early warning of attack:	Protect Arctic forward EW stations:	Vojska PVO
2.2 Prevent LRB reaching targets:	Establish Arctic/Nordic air defence intercept barriers	Vojska PVO
2.3 ICBM early warning of attack:	Protect Arctic forward EW stations:	Vojska PVO
2.4 (Prevent ICBMs reaching targets?:	Protect Arctic ABM target acquisition radars:	Vojska PVO)
2.5 MRB:	As for long-range bombers:	Vojska PVO
3. Strategic sea defence		
3.1 Prevent SSBN launch of SLBMS:	Prevent MR SLBM SSBNs reaching northern launch areas:	Navy
	Destroy those SSBNs which are in Northern Waters:.	Navy
	(Track and destroy LR SLBM SSBNs in distant oceans?:	Navy)
3.2 Prevent SSGN launch of SLCMs:	As for MR SLBM SSBNs:	Navy
4. Strategic naval conventional		
4.1 Establish CENTLANT presence (peacetime):	Build up and deploy large Northern/Baltic Fleet:	Navy
4.2 Cut Atlantic SLOC (wartime)	Attack submarine and air penetration of NATO NORLANT ASW/AD barriers:	Navy
4.3 Maintain fleet in Third World waters (peacetime):	Build up and deploy large Northern/Baltic Fleet:	Navy

Table 6.3: The translation from regional strategic support requirements to specific northern TVD strategic support missions

Northern strategic support required	Northern TVD strategic support missions	Service with primary responsibility
1. Strategic nuclear offensive		
1.1 Protect Arctic forward airfields:	Long-range Arctic air intercept barriers; Short-range Nordic air intercept barriers; Point air defences of major targets; SR/MR naval barriers (as 1.4):	Vojska PVO: IAPVO,Vojska RPV Vojska PVO: IAPVO/Vojska RPV Vojska PVO: Vojska ZRV/Vojska RPV Navy
1.2 Provide Arctic long-range escort:	Long-range Arctic fighter escort:	Vojska PVO: IAPVO/Vojska RPV
1.3 Protect Northern Fleet Arctic SSBN bases:	As 1.1:	Vojska PVO/Navy (as 1.1)
1.4 Protect LR SLBM Arctic launch areas:	Forward sea denial ('Outer Zone of defence'); SR/MR sea control ('Inner Zone of Defence') Complete local sea control ('Ocean Bastions'):	Navy Navy Navy
1.5 Assist MR SLBM SSBNs reaching WESTLANT launch areas:	Local and temporary forward sea denial ('Outer Zone of Defence'):	Navy
2. Strategic air defence		
2.1 Protect Arctic forward EW stations:	As 1.1:	Vojska PVO/Navy (as 1.1)
2.2 Establish Arctic/Nordic air defence intercept barriers:	As 1.1 except for navy missions:	Vojska PVO
2.3 Protect Arctic forward EW stations:	As 1.1:	Vojska PVO/Navy (as 1.1)
2.4 (Protect Arctic ABM target acquisition radars:	As 1.1:	Vojska PVO/Navy (as 1.1))
2.5 MR bombers as for LRB:	As 1.1 except for navy missions:	Vojska PVO
3. Strategic sea defence		
3.1 Prevent MR SLBM SSBNs reaching northern launch areas:	Sustained forward sea denial ('Outer Zone of Defence'):	Navy
3.2 Destroy those SSBNs which are in Northern Waters:	Temporary MR and SR sea control ('Inner Zone of defence'):	Navy
3.3 (Track and destroy LR SLBM SSBNs in distant oceans?:	ASW SSN penetration of NATO NORLANT ASW barriers; As 1.5:	Navy)
3.4 SSGN as for MR SLBM SSBNs:	As 3.1 and 3.2:	Navy
4. Strategic naval conventional		
4.1 Attack submarines/air penetration of NATO NORLANT ASW/AD barriers	As 1.5 and 3.3:	Navy

Strategic mission	*Force responsible*
1. Sea-launched strategic nuclear attack:	VMF (SSBN forces)
2. Defence against seaborne nuclear threat:	VMF (Strategic ASW forces)
3. Defence against airborne nuclear threat:	Vojska PVO
4. Airborne strategic nuclear attack:	AASU (36. AASU and 46. ASSU)
5. Cutting NATO Atlantic SLOC:	VMF (sea denial forces)

These primary Soviet strategic interests have no direct military relevance for the Nordic area, and the forces assigned to this region for their implementation do not, with two exceptions, directly affect the regional military equilibrium of power. The two exceptions are the Vojska PVO and the VMF sea-denial forces, whose regional order of battle does affect the regional equilibrium of power, and whose forces, when not applied to their primary strategic task, could affect the outcome of a conventional military confrontation in the north. In such a case, they are no longer acting in their strategic role but in their secondary regional theatre role. As such, they will be examined below.

However, indirectly, the Soviet strategic interests in the north-west affect the regional military situation. This is because each of the forces listed above requires a certain number of additional military forces and wartime operations to safeguard their security. It is these operations, and the forces and preparations in peacetime necessary for their wartime implementation, which directly affect the situation.

Figure 6.10 shows the Soviet military perception of the Northern Waters area. This is divided into one major land theatre of operations, the North-western TVD, and two major ocean theatres of operations, the Arctic and Atlantic OTVDs. The relationship between the major Soviet strategic interests in this area and her regional theatre-level military operational interests will now be examined in two stages. Firstly, the translation of major Soviet northern military strategic interest into northern theatre-level strategic military support missions will be outlined. Secondly, the translation of these main strategic support missions into the additional regional theatre and front level ground operations which are necessary for their success will be recounted.

THE TRANSLATION OF SOVIET NORTH-WESTERN STRATEGIC INTERESTS INTO NORTHERN TVD STRATEGIC SUPPORT MISSIONS

Tables 6.1, 6.2 and 6.3 outline the process whereby the primary Soviet northern strategic objectives are translated into the first stage of their regional support requirements. Essentially, this consists of an outline on Table 6.2, firstly of the regional strategic support requirements which are necessary for the implementation of the strategic missions, and secondly (in Table 6.3) of the additional northern support missions which are required if the strategic support requirements are to be fulfilled. It is the northern strategic support requirements which are of interest here, as they start to affect directly the northern military situation. A closer look at these shows that they essentially consist of two main types of mission, each composed of a certain number of subordinate missions. These are:

Main mission	*Sub-missions*	*Force responsible*
Air Defence	Long-range intercept	IAPVO, Vojska RPV
	Medium-range intercept	IAPVO, Vojska RPV
	Short-range intercept	IAPVO, Vojska RPV
	Area and point defence	Vojska ZRV, Vojska RPV
Naval Defence	Forward Sea Denial: 'Outer Zone of Defence'	VMF sea denial forces
	Inner sea control: 'Inner Zone of Defence'	VMF sea control forces
	Complete local sea control: 'Arctic Ocean Bastions'	VMF sea control forces

Thus, there are essentially two major strategic support operations required by the USSR in the Nordic region and two services, the Vojska PVO and the VMF, with primary responsibility for their execution. Both of their strategic support missions are important for the safety and operational viability of the Soviet strategic interests and forces in the north-west. The first of these protects most of the vital regional strategic facilities in the area, more specifically in the North-western TVD and Arctic OTVD, from air attack. These include, in an approximate order or priority: the SSBN bases in the Leningrad Military District (MD) and the SSBN forces on station in the Arctic

OTVD; the BMEW radars along the Arctic coast in the Leningrad MD; and the protection of the forward air-bases in the Leningrad MD for the AASU. The geographical scope of these air defence missions is outlined in Figure 6.11. It is important to note that in this case these missions overlap with the main strategic missions of this service, including that of guarding the major bomber and ALCM transit routes which pass over this area.

The second set of missions protects roughly the same strategic facilities, but this time from sea-launched attack. However, a major priority is placed upon the protection of the SSBN forces in their concealment and launch stations in the Arctic. Since the early 1970s this mission most likely has constituted the single most important *raison d'être* of the Northern Fleet, and the forces assigned this task probably include most of the major surface and submarine combatants of this Fleet, and elements from the Baltic Fleet as well, though the exact distribution of sea-denial forces (attack submarines and anti-shipping aviation of the MA) between offensive central Atlantic SLOC interdiction and strategic defensive GIFUK sea-denial barrier operations is uncertain, and would vary depending on the scenario.

Figure 6.12 shows the approximate geographic scope of this strategic support mission. It primarily involves the Atlantic OTVD, and includes three main operations in this area. Firstly, there is the establishment of an Outer Zone of Defence commencing roughly along the GIFUK Gap and involving primarily sea-denial operations against surface combatants using submarines and naval attack aviation. Secondly, there is the establishment of an Inner Zone of Defence, with its outer perimeter roughly along the central Norwegian Sea. Operations in this area involve the attempt to establish a degree of sea control, permitting the major surface combatants to carry out strategic ASW against Western attack submarines which have entered the area and which could threaten either shore installations or, more importantly, could carry out ASW operations against the Soviet SSBN strategic nuclear reserve in its Arctic sanctuaries. Finally, the establishment of Ocean Bastions around the main operating areas of the SSBN forces includes the Barents, Greenland and Kara Seas, as well as the Arctic Ocean, which itself is probably a major operating area for the *Typhoon* SSBN.

It is quite clear that these missions, which still have as their primary objective the support of the strategic forces, also drastically affect the regional northern military situation. This is because the nature of their strategic support role, essentially consisting of controlling the skies for the Vojska PVO and controlling the sea for the VMF, also affects the

Figure 6.11: Soviet northern air defence operations

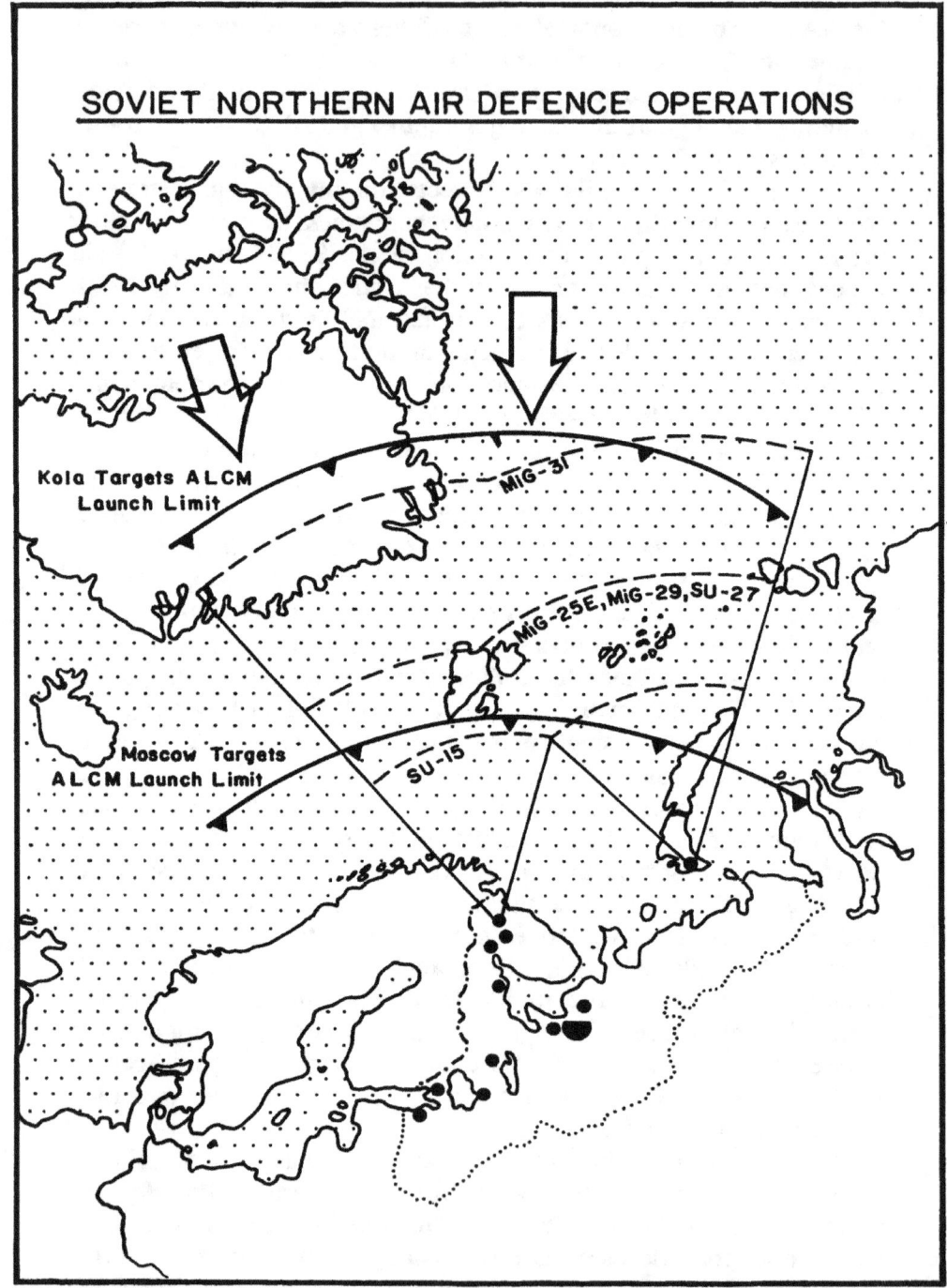

Figure 6.12: Soviet northern naval defence operations

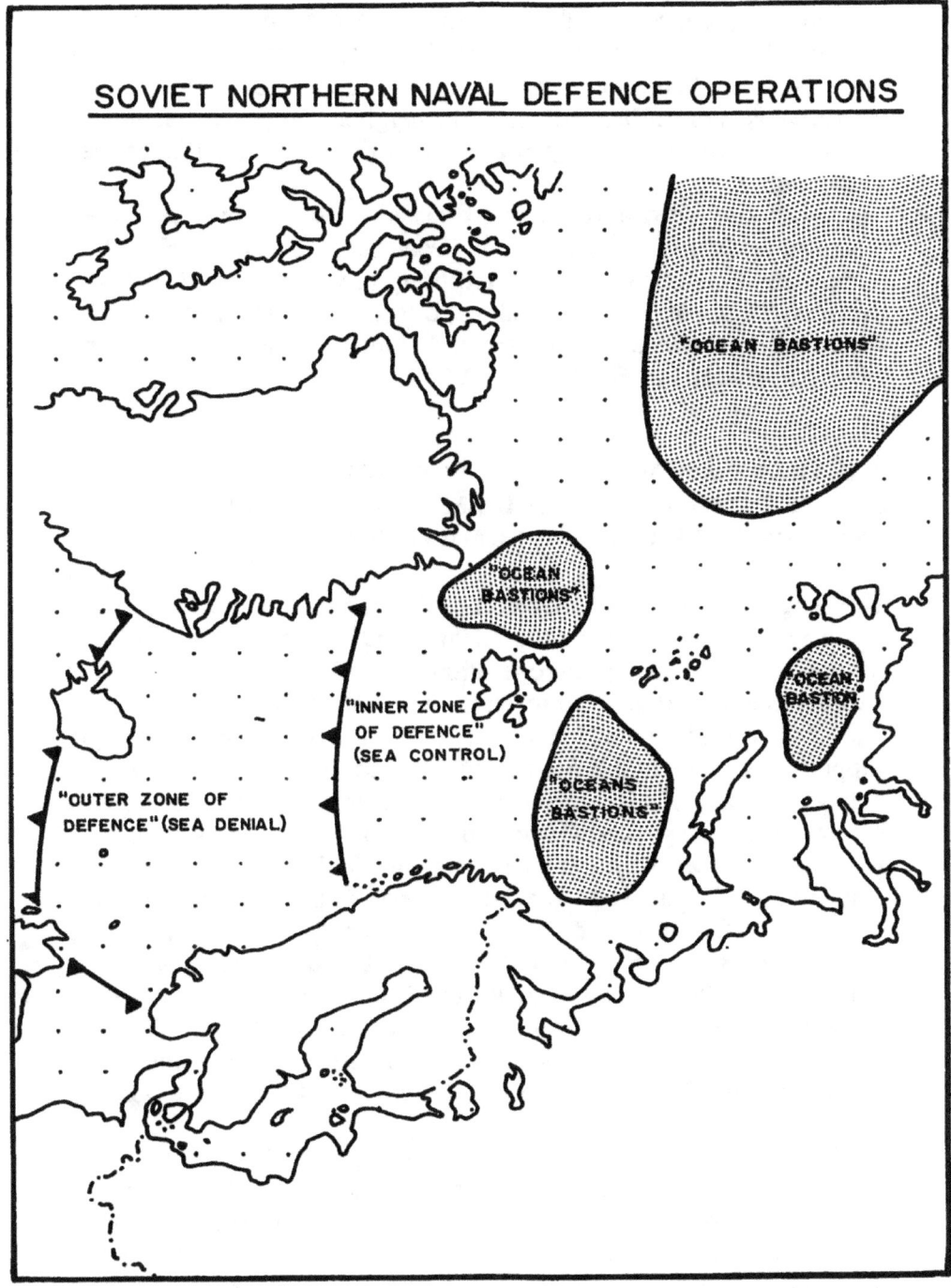

regional military equilibrium of power. As the range of Soviet air control and the sophistication of the forces assigned to it grow, the corresponding capacity of the adjacent states to control their airspace declines. The same applies to the naval forces, which, as their control of the Northern Waters expands southwards, reduce the credibility of existing levels of military forces in the neighbouring states leading to a need for greater military forces and/or increased Alliance ties westwards.

In addition, the strategic support missions listed above also affect the northern military equilibrium because, to be successful, they require additional Soviet regional military preparations and forces in the area. This leads to a second stage in the translation of military requirements down to the theatre and front level, mainly involving ground operations in the North-west TVD.

THE TRANSLATION OF SOVIET NORTH-WESTERN STRATEGIC SUPPORT REQUIREMENTS INTO REGIONAL THEATRE-AND FRONT-LEVEL MILITARY OBJECTIVES

This stage involves the translation of the two main strategic support missions in the Arctic and Atlantic OTVDs into subsidiary theatre-level ground support operations in the North-west TVD. This requirement for further military support operations on the theatre level arises because of two basic factors. Firstly, the Soviet naval strategic support operations in the Atlantic OTVD require air cover if they are to have a chance of succeeding. This air cover is presently land-based in areas which are so far removed from the main wartime naval combat zones as to make its air defence contribution difficult. Therefore the land invasion of areas in the North-west TVD favourable for the operation of forward air defence forces is vital. Secondly, the close geographic proximity of a number of the vital North-west TVD strategic facilities on the Kola to the NATO command area in northern Norway makes these facilities, at least in the eyes of Soviet planners, vulnerable to surveillance and air attack. Thus, a land invasion of the bordering NATO areas is likely to be perceived as a vital wartime operation.

Of these two requirements, the one of acquiring forward air bases is particularly important. The Northern Fleet would be facing a tremendous concentration of air power in the northern Atlantic, comprising the very powerful US Navy aviation, including air defence/superiority, ASW and anti-shipping forces operating from carriers and shore support facilities, as well as large numbers of NATO land-based air defence/superiority, anti-shipping and Airborne Warning and Control (AWACS) forces, operating mainly from Iceland and Scotland, as well as from a number of further adjacent coastal areas. Without air cover it would be very difficult, probably impossible, to maintain a surface presence in the Inner Zone of Defence, and, in the long run, to maintain the sea denial forces in the Outer Zone of Defence. The former would be subjected to devastating air attack, while the latter would be subjected to heavy airborne ASW operations (against the SSN/SSGN) and air defence barriers (against the naval anti-shipping aviation). As a result air cover is vital to the Northern Fleet operations in wartime.

For the moment the only such air cover available is from the long-range interceptor forces of the IA PVO based in the Arkhangelsk ADS. This has been the case since the transfer of MA air defence forces to the PVO Strany in 1956 and will remain so until the deployment of the navy's first true carrier-based air defence forces in sufficient numbers to make them able to stand up to the Western naval air superiority forces. The first of these carriers is under construction in the Black Sea and the Soviets are presently trying out the Su-27 *Flanker*, the MiG-29 *Fulcrum* (both primarily for air defence/air superiority) and the Su-25 *Frogfoot* (attack) aircraft for use from the carrier. While these forces may become operationally available to the VMF by the 1990s, it will be some time before sufficiently large carrier-based forces, with the requisite know-how, are available to counter Western air forces in the North Atlantic. Until this time they will continue to remain dependent upon air support from ground-based IA PVO forces.

Thus, a major additional task for the Vojska PVO is to provide air cover for the navy in the Atlantic OTVD. Since at least the mid-1970s and probably for quite some time before that, it has carried out exercises in the area to practise this mission, mainly involving long-range fighters operating in conjunction with the Tu-126 *Moss* long-range forward AEW aircraft. However, if it is to provide adequate wartime cover for the fleet, it is vitally dependent upon acquiring forward air bases as the present bases in the Arkhangelsk ADS are too far removed from the main Atlantic combat zones and leave the IA PVO forces with virtually no time on station in these areas.

Such potential forward air bases exist in the form of airfields in northern Finland (though these are still fairly remote), throughout the length of Sweden, and in all of Norway. The advantages to the Vojska

Figure 6.13: Air control using only Soviet bases

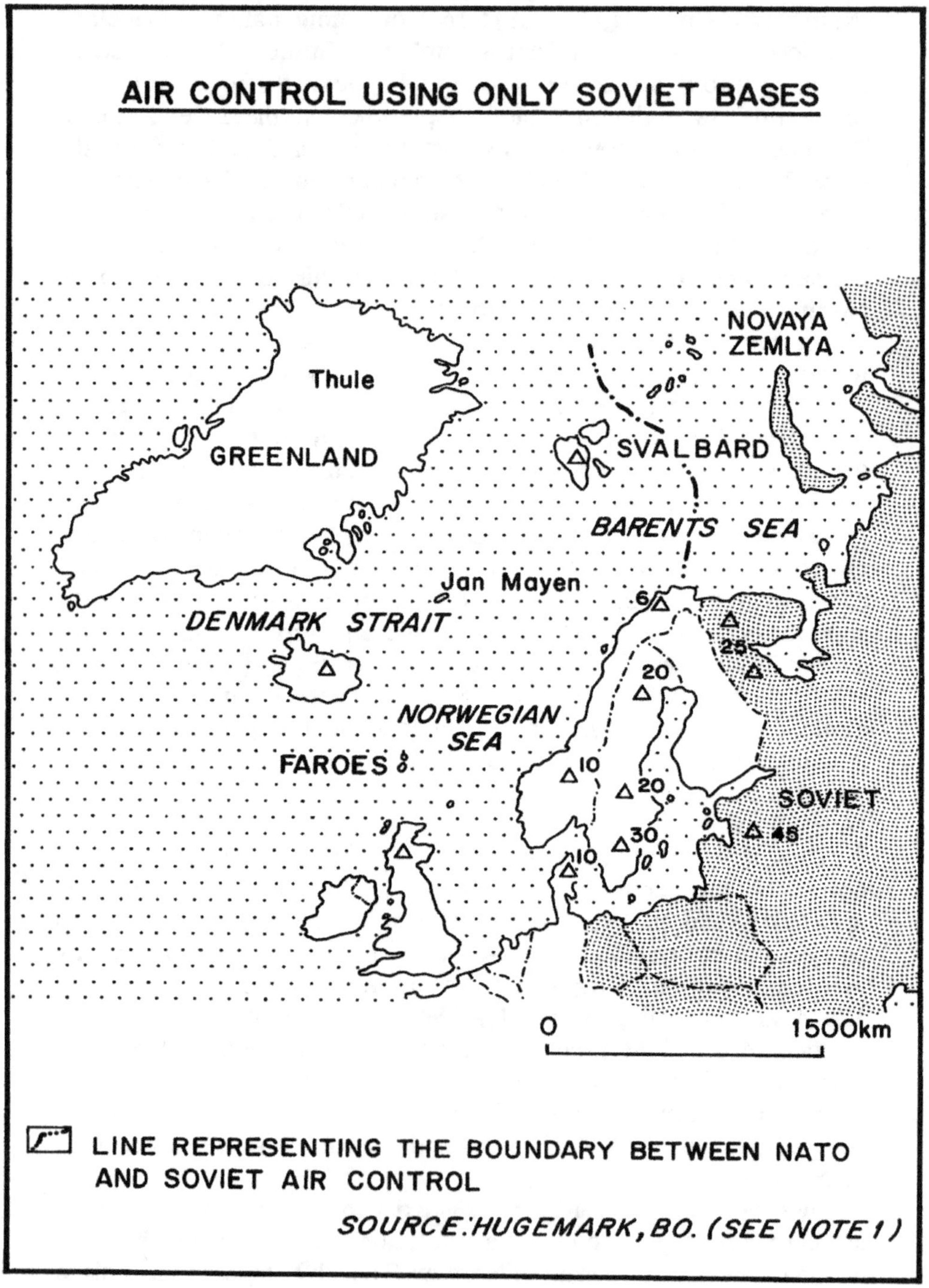

Figure 6.14: Air control using North Norwegian bases

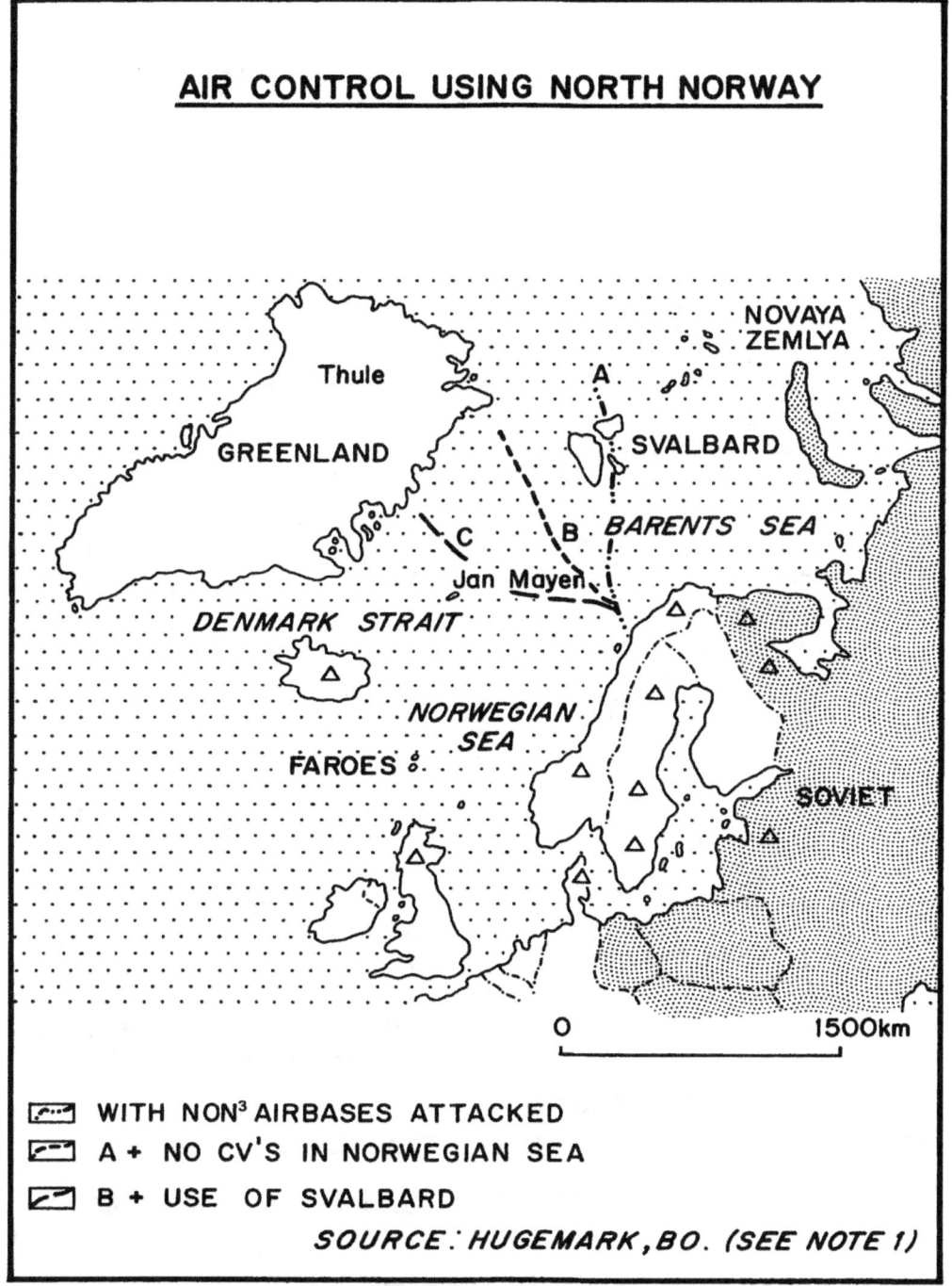

Figure 6.15: Air control using North Norwegian Sea

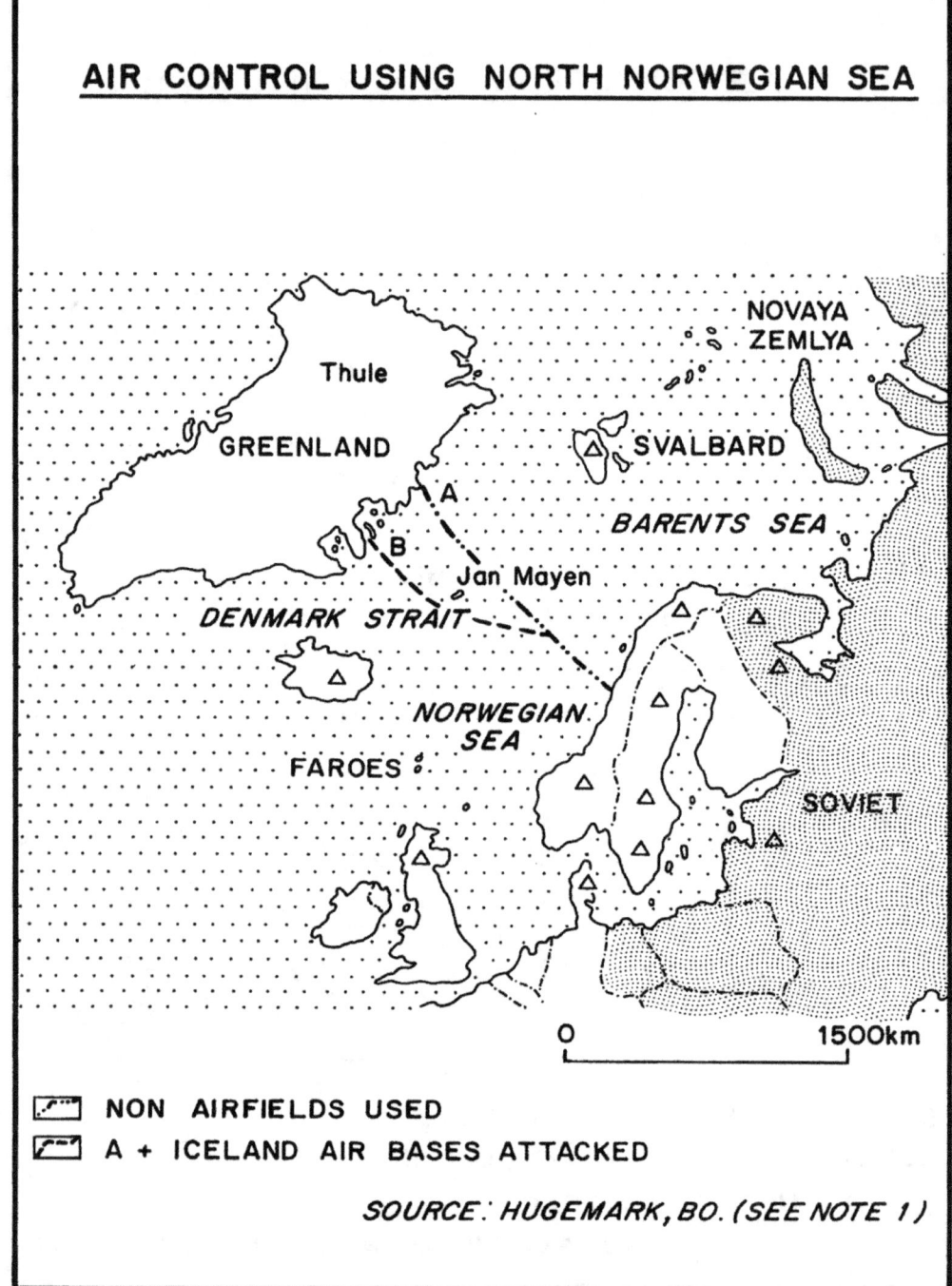

Figure 6.16: Air control after attack on Norwegian bases

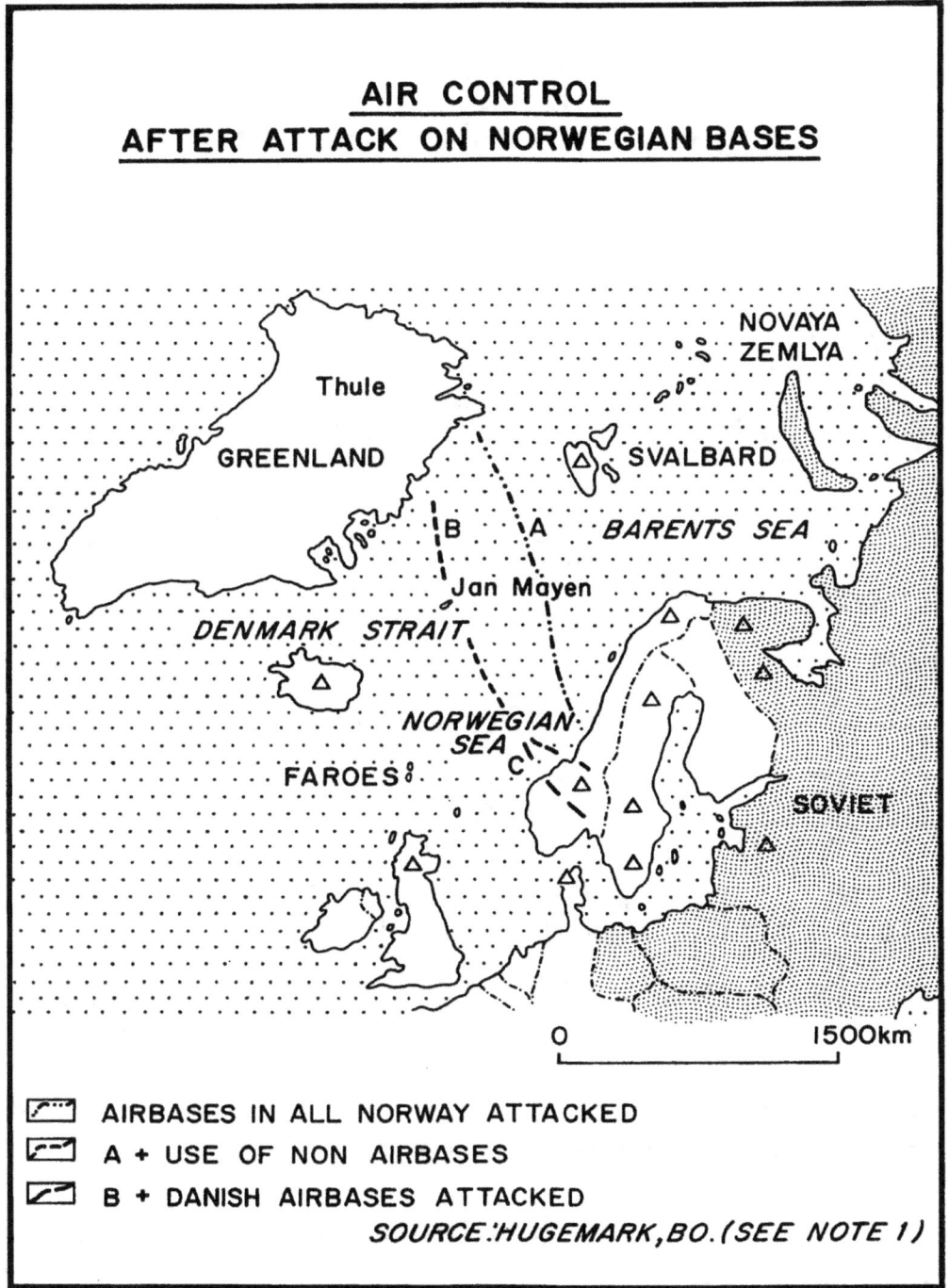

MILITARY STRATEGY

Figure 6.17: Air control using Scandinavian bases

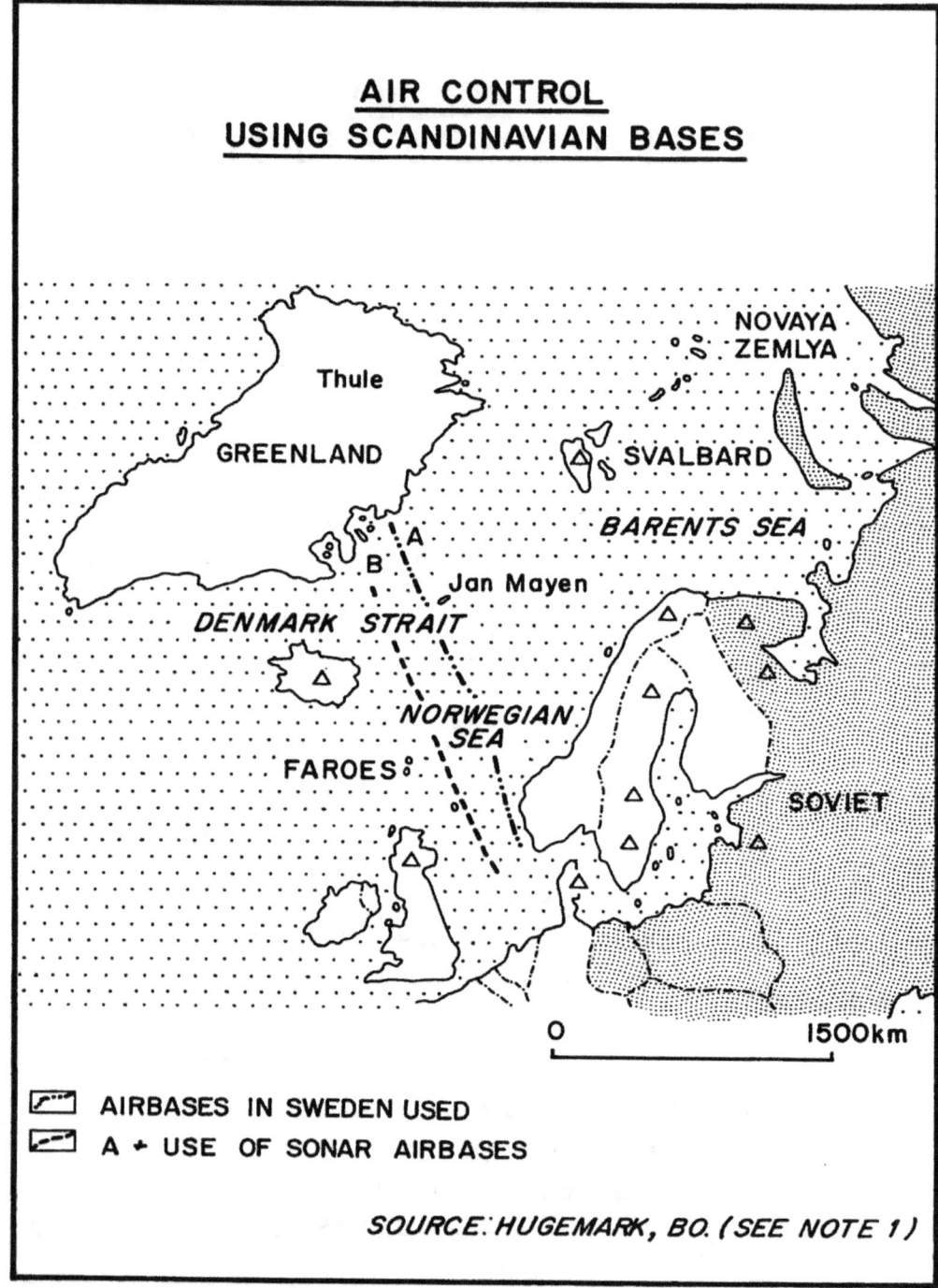

Figure 6.18: Probable main thrust of Soviet ground operations in wartime in the North-western TVD

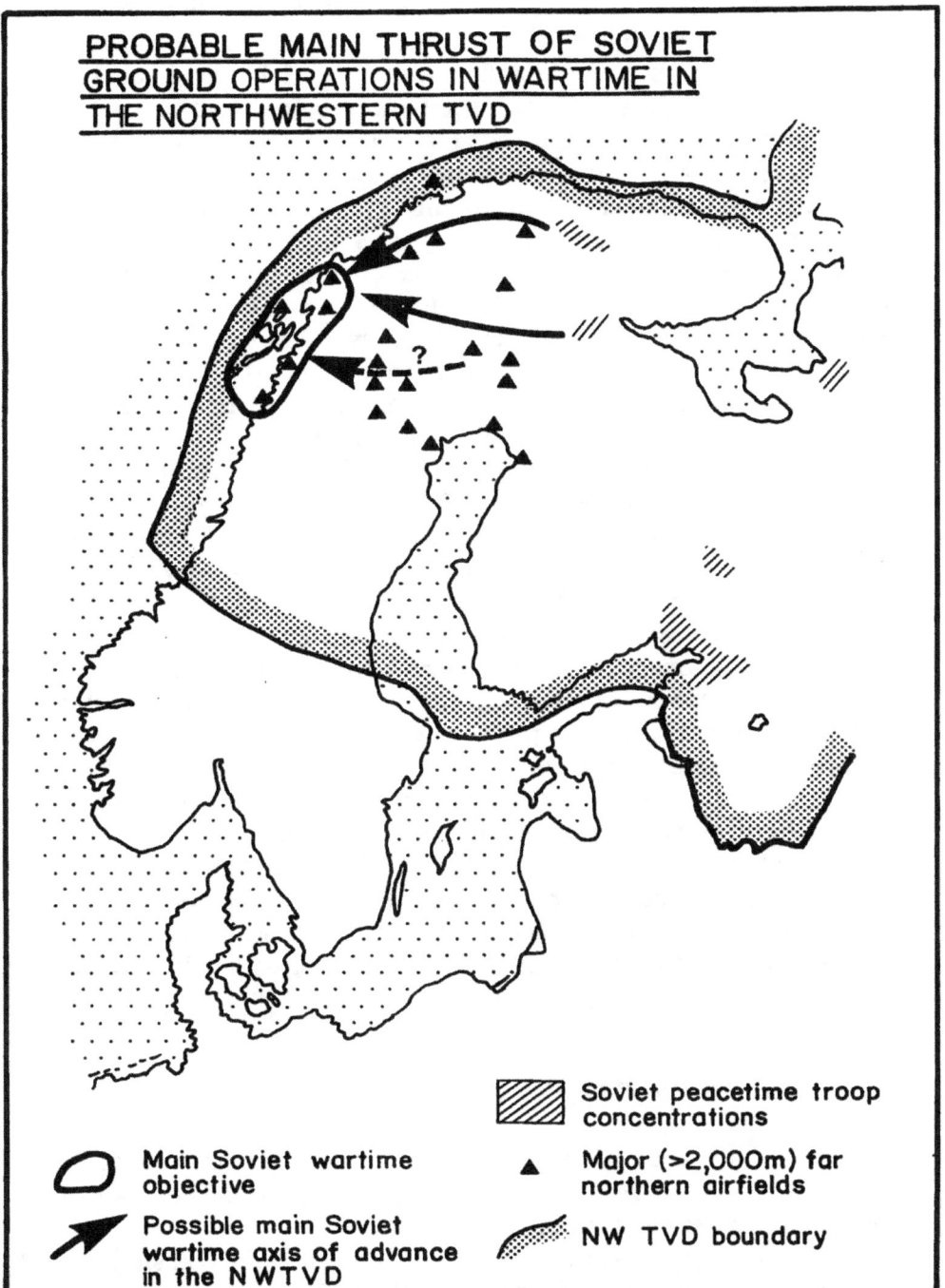

PVO air defence effort in the Atlantic of using a combination of these bases has been calculated and demonstrated by Lt. Colonel Bo Hugemark[1] of the Swedish armed forces, whose maps (Figures 6.13 to 6.17) are reproduced here. These illustrate the considerable advantages which utilisation of these forward airfields would confer.

The wartime operation of these airfields requires their occupation and protection, which in turn calls for ground operations to seize the relevant territory. As a result we may expect the Soviet armed forces to make a third major conventional military effort in the region in the event of war, this time involving the ground operations necessary for obtaining forward air bases. Such an operation would probably concentrate on northern Fenno-Scandinavia. This is because these air bases would have to be occupied very rapidly if they were to be of any use to the IA PVO, and it is in the far north where the regional balance of forces is the most favourable to the USSR, and where the distance between the USSR and the airbases is the shortest.

These operations would primarily be the responsibility of the commander of the NW TVD. Figure 6.18 shows the approximate scope of the initial ground operations and most of the airbases in northern Fenno-Scandinavia. The operations would probably involve ground advances across northern Norway and northern Finland for the key objective of the Troms area in northern Norway. In addition, they could involve amphibious assault and airborne assault operations and would probably include very heavy air attacks. It is important to keep in mind that these ground operations probably constitute a third order priority in the regional strategic planning, as they essentially consist of support operations for the Vojska PVO support operations for the primary Soviet north-western strategic interests. As such there has been considerable care on the part of the USSR not to build up the conventional ground and Frontal Aviation forces in the northern part of the Leningrad MD in peacetime, and there may well have been an effort on the part of the USSR not to increase tensions in this area unnecessarily.

THE CONSEQUENCES FOR REGIONAL STABILITY

Soviet military strategy in the north-western neighbourhood of the USSR can be subdivided into three major categories. These are, in descending order of Soviet strategic priority:

1. Four major existing strategic interests (SSBN basing and operations; Strategic air defence; Strategic ASW; Long-range bomber

support) and one major emerging strategic interest, that of cutting the North American–European trans-Atlantic SLOC both physically (in wartime) and psychologically (in peacetime).

2. Two main strategic regional support requirements to ensure the operational viability of the above primary interests and the physical safety of the forces assigned to their implementation: the theatre-level air defence of the above strategic assets and their basing and operating stations in the north-west; the theatre-level sea defence of the Arctic SSBN Bastions and the Kola basing infrastructure.

3. One main regional subordinate strategic support requirement to assist the forces assigned the main regional strategic support missions in the execution of their duties: theatre-level ground offensive in northern Fenno-Scandinavia to secure the vital airfields in northern Norway and possibly Finland and Sweden.

The consequences of each of these for the Northern Waters region will be examined in turn.

The first of these categories does not directly affect the regional military equilibrium of forces, but decisively affects the overall stability of the area by determining the level of Soviet (and US) military-strategic interest in the area. This has two main consequences: firstly, it helps determine the level of Soviet regional military presence and investment in the area; and secondly, if it results in the attraction of US military attention and forces, it increases the level of military tension in the region. On this level one can make three main conclusions:

- The north-western region (the Arctic and Atlantic OTVDs and the North-western TVD) does not in itself constitute a primary strategic objective with an intrinsic strategic value, the control of which would significantly alter the global correlation of forces in favour of the Soviet Union. Therefore this area is not likely to be the main target of a major military confrontation, nor is an isolated north-western Soviet military operation, designed only to accomplish regional goals, likely under present and foreseeable political circumstances.
- However, the north-western region is of major indirect military strategic importance to the USSR as a basing and deployment area for four (and soon perhaps five) vital strategic forces. The importance of these for the USSR has, in a majority of significant cases, grown over the last three decades, significantly augmenting Soviet military strategic interests in the area.
- Soviet interest in the area is likely to increase because of the deployment of the *Typhoon* SSBN, vastly increasing the strategic

importance of the Arctic Ocean to the USSR; the development of the SDI possibly leading to the deployment to the north-west of long-range bombers and SSBNs armed with short-range SLBMs and the Arctic coast being used for Soviet SDI efforts; the modernisation of US strategic bomber forces, increasing the air defence value of the north-west; the modernisation of Soviet strategic bomber forces which could use the Arctic coast for basing and support; the deployment of the US SLCM renewing the need for strategic ASW in the Norwegian Sea; and the growing Soviet fleet which increases the USSR's perception of her own importance and influence in Northern Waters and Scandinavia.

For the local states all this implies a deterioration of the strategic environment, as it reduces the capacity of the regional states to avoid being drawn into superpower tensions and could lead to increased peacetime superpower confrontations in the area itself, notably in the Arctic, where the capacity of the regional states to moderate and act as buffers is at the lowest.

This situation has already led to one major reaction on the part of the Nordic states, in the form of the increased calls in some circles for the creation of a Nordic Nuclear-Free Zone (NNFZ) which is clearly an attempt to try to deal with one of the potential future main sources of regional tension. Perhaps more attention should be given to the suggestion made by Dr Willy Østreng in 1982 for an Arctic SSBN sanctuary.[2] This area is a far more likely source of regional trouble than that which is presently covered by the NNFZ proposals.

The second of the main categories of Soviet military strategic interest in the north-west — the need for strategic military support operations to safeguard the above main strategic interests and forces — affects the regional equilibrium of military power by creating a series of regional theatre military forces which are assigned regional theatre-level strategic support missions. However, while the forces assigned this task — the conventional sea-control and sea-denial units of the VMF and the fighter-interceptors and associated forces of the Vojska PVO — directly affect the regional equilibrium of military power, shifting it in favour of the USSR, their primary orientation is towards defence against the major threat to their regional strategic interests. As this derives from extra-regional powers, it moderates their direct military influence on the regional states.

On this level one main conclusion can be drawn. Because of the gradual growth of the importance of the main Soviet strategic interests in the north-west, the secondary importance of the support forces has also grown over the last 30 years. This has led to a conscious and major Soviet effort to build up its conventional naval and air defence forces in the area, to considerable increases in their regional order of battle, and to a very significant qualitative increase in their capabilities. In the coming years, this trend is likely to continue due to the continued growth of the strategic importance of the north-west.

The first overt reaction of the regional states to this situation came in 1981, with the Norwegian–US POMCUS agreement for the prepositioning of some heavy equipment for one US Marine Amphibious Brigade (MAB) in central Norway. On the one hand the agreement is the direct consequence of the growing Soviet naval might in the Norwegian Sea, which endangered the credibility of the vital NATO capacity for reinforcing Norway by sea. On the other hand the propositioning depots were placed in central Norway, relatively far from the operational area in northern Norway. This in turn reflects the political dimension of security policy, incorporating the equally vital need not to raise regional tensions *vis-à-vis* the USSR more than necessary. This dilemma may continue as the NATO Nordic states have to choose between maintaining either their military deterrent credibility or regional good relations with the USSR. This reassurance–deterrence dilemma will mainly affect the problem of balancing the two major Soviet regional strategic support requirements, and will not significantly affect the 'higher' level of the main regional Soviet strategic interests, nor the 'lower' level of the Soviet theatre-oriented military support operations for the strategic support operations.

The third of the levels of Soviet military strategic interest in the north-west directly affects the regional equilibrium of military power and the operational orientation of the forces. This level involves the Soviet need for theatre-level military operations to provide support for the naval and air defence strategic support forces. This primarily involves ground operations in the northern part of the NW TVD, directed towards the seizure of the vital airfields in this area, exploitation of which is crucial for the success of the naval defences in the Norwegian Sea.

The growing importance of this naval strategic support mission over the last three decades has increased the importance of this final level of operations. However, for various reasons it has not led to an immediately evident increase in the relative strength of Soviet ground forces deployed to the Leningrad MD. While these forces have undergone considerable modernisation and their theatre-level and tactical mobility has notably increased, this process has on the whole not been greater than the modernisation of the total Soviet ground and frontal aviation forces. Thus, the regional order of battle of this level of forces has not increased significantly numerically, and the qualitative increase is similar to that generally experienced by the Soviet armed forces. Indeed it seems that the USSR may have made deliberate efforts not to increase her Kola-based tactical forces to a level which could alarm her neighbours. There would therefore seem to be some potential for maintaining a rough degree of mutual restraint on this level. Nonetheless, the evolution of Soviet tactical forces does require a constant updating of the Nordic national defence capabilities, though this is substantially different from matching Soviet regional strategic support forces. Maintaining a roughly credible deterrent and defence capability against Soviet northern tactical military developments is a matter which each of the Nordic states, in some cases with some sacrifice, can do individually.

CONCLUSIONS

Soviet military strategic interests in the north-west can be divided into the three levels presented here. The increase in Soviet interests and deployment of regional forces in the first two levels is leading to problems for the regional states in preserving the strategic stability of their environment. However, the problem on each of the three levels must be approached in different ways. The highest level probably calls for a primarily diplomatic, multilateral arms control or confidence-building approach intended to reduce superpower regional antagonisms. The most important area to focus on here is the Arctic, as this is a primary source of Soviet regional strategic interest. Under the prevailing East–West political climate the chances for success here seem slim.

The second level may offer better opportunities for resolution but also includes the major Nordic dilemma of how to balance reassurance with deterrence. Resuscitating the possibilities for regional stability on this level probably requires a mix between unilateral regional military adjustments and attempts to fit them in with a regional confidence-building approach. For Norway this would have to include the approval — at the very least — of her major NATO allies. The fact that this area is more sharply regionally focused may increase the scope for arriving at limited compromises suitable for both the US and the USSR. This level could also involve unilateral attempts by the regional neutral states to maintain or build up regional military confidence-reinforcing measures, such as the vital Swedish air defence barrier capability. As this is an expensive financial burden for one state, it could alternatively call for a common Nordic neutral military approach, for instance through the creation of a united 'Finnish-Swedish Lappland Neutral Air Defence Zone', though the political problems associated with this approach may prove insurmountable. On a military level this makes sense, as it could both reduce Soviet fears of the cruise missile threat in this area and reduce NATO fears of Soviet air strikes over this area. Finally, of course, it could reduce the air defence costs of both Finland and Sweden through the benefits of the common venture. It would, however, depend on Soviet political confidence in the neutrals.

The developments on the third level have been the least destabilising and have involved relatively the smallest Soviet force increases in the NW TVD. Regional stability at this level can probably best be met by unilateral military force adjustments by each Nordic state individually, and should not involve any major conceptual dilemmas, though it would naturally impose a financial burden on each state.

NOTES

1. Bo Hugemark, 'Nordatlanten och Norden - en krigskådeplats', *Kungliga Krigsvetenskapsakademiens Tidskrift*, vol. 186, no. 1 (1982), pp.31–43.

2. Willy Østreng, 'Strategic developments in the Norwegian and polar seas: problems of denuclearization', *Bulletin of Peace Proposals*, vol. 17, no. 2 (1982), pp. 101–12. Willy Østreng, *Sovjet i Nordlige Farvann*, 1st edn (Gyldendal Norsk Forlag, Oslo, 1982).

3. NON = North Norway.

7

A Naval Force Comparison in Northern and Atlantic Waters

Robert van Tol

INTRODUCTION

The strength of the Soviet Navy in Northern Waters is frequently exaggerated by the simple assessments that are common in the public discussion of the naval balance. A more detailed analysis reveals that the Soviet Northern Fleet has serious limitations and problems, in particular the possession of large numbers of either small, aged, or technologically unsophisticated vessels. Given these limitations, it seems unlikely that the Soviet Navy could mount any sort of large-scale assault on NATO's sea lines of communication (SLOC) without seriously weakening its own ballistic missile submarine (SSBN) bastion. NATO, on the other hand, can bring to bear a much larger, more modern and more capable force in the Northern Waters, but faces the problem that these forces are distributed across a far-flung basing structure and do not normally operate in the hostile environment of Northern Waters in significant numbers.

Despite the inhospitability of the environment, the Soviet Union makes extensive use of its Arctic territories and their adjacent Northern Waters for military purposes. In particular, this region plays a significant part in the Soviet Union's strategic nuclear offensive and defensive forces, with missile test sites, anti-ballistic missile radar sites, advanced air bases for strategic bombers and other such facilities. However, the most important element of the Soviet Union's strategic nuclear forces deployed in the Arctic region is the Soviet Navy's force of nuclear-powered SSBNs.

The Soviet Navy's SSBNs are deployed from two main bases. Some 40 per cent of the SSBNs are based at Petropavlovsk on the Kamchatka Peninsula in the far North Pacific, just south of the Aleutian Islands and the Bering Sea. Although these Pacific-based SSBNs could deploy into Arctic waters through the Bering Strait, their main orientation is towards the central Pacific basin and so they are not dealt with here. The other 60 per cent of SSBNs are based on the Kola Peninsula with the Soviet Northern Fleet. This Fleet is the largest and most important of the Soviet Union's three European fleets, and is roughly equal in size, though more modern, than the entire Pacific Fleet. It numbers about 500 vessels[1] and 400 naval aircraft,[2] conducting extensive operations throughout Northern Waters.[3] It is this force which will be compared with its potential NATO opposition from a number of standpoints. Such evaluations will exclude the land–air balance of forces in Finnmark, but will include the overall size, distribution[4] and age of the opposing forces. The detailed calculations to the following discussion may be found in the Appendix to this chapter.

Size: The Northern Fleet vs NATO in Northern Waters

Gauging the size of the respective naval forces which could operate in Northern Waters is not a matter of simple accounting. The exact level of forces which would operate in this region is subject to specific conditions which are impossible to predict.[5] Thus, any simple numerical comparison is only an estimate based on a series of assumptions.

Table 7.1: Soviet Northern Fleet/NATO's Atlantic naval forces: the aggregate statistics[6]

	Soviet forces	NATO forces
Ballistic missile submarines	42	31
Attack submarines	132	136
Combat aircraft	520	1,050
Open ocean warships	42	206
Coastal warships	95	66
Mine countermeasures vessels	65	49
Assault ship (displacement tonnage)	38,050	402,300
Marines	3,000	56,250
Coastal defence troops	2,200	1,000

Note: NATO's Marines include a US Marine Division.
Sources: Abstracted from Appendix Tables 7A-1, 7A-2 and 7A-3.

The size of the Soviet Northern Fleet, being geographically isolated from other Soviet naval forces, is fairly easy to tabulate. The only major reinforcement to its strength presented here (see Table 7.1) comes from non-naval aircraft based on the Kola Peninsula. Nonetheless, these figures should be considered as illustrative rather than as exact. Air power, for example, by its nature is highly flexible and its bases can readily be changed.[7] Likewise, ships and submarines of either the Baltic Fleet or the Black Sea Fleet could operate with the Northern Fleet, though to any significant extent this is unlikely since their access to the Atlantic is restricted by choke points and they themselves have demanding tasks within the Baltic, Black and Mediterranean Seas.[8]

Estimating the size of NATO naval resources available for northern operations is substantially more difficult. NATO's chief naval commander, the Supreme Allied Commander Atlantic (SACLANT), has two main responsibilities. He must ensure the security of the reinforcement shipping from America to Europe, along the SLOC; and he must provide support to the Supreme Allied Commander Europe in both the Southern and Northern Flanks of NATO, either of which could theoretically divert resources away from SLOC protection. Protecting the SLOC, however, is SACLANT's first priority mission, and as such it will keep a large proportion of his resources both out of the Mediterranean and south of the Greenland–Iceland–UK (GIUK) Gap.[9] The NATO forces in Table 7.1 represent a gross force pool available to SACLANT for operations in the Atlantic or on the flanks. It is drawn from the naval forces of the NATO members on the Atlantic coast and neither assumes any diversion of units to the flanks (such as extra Carrier Battle Groups to the US 6th Fleet in the Mediterranean), nor any strengthening of forces in the Atlantic (such as reorientating the primarily Mediterranean-based French fleet or transferring 6th Fleet carriers to the Atlantic US 2nd Fleet).[10] The level of uncertainty in these gross figures should be borne in mind in the following discussion.

DISTRIBUTION BY BASING AREAS

The Northern Fleet's bases are highly concentrated on the Kola Peninsula. With the exception of the Northern Fleet's main shipyard at Severodvinsk and the commercial harbour at Arkhangelsk (both on the White Sea coast), these bases are clustered around the Kola Inlet only 50 km away from the Norwegian border.[11] This makes them vulnerable to attack and surveillance but allows for easy concentration of force. NATO's position is the exact opposite, with a highly dispersed network of bases stretching from North Cape to Gibraltar and across the other side of the Atlantic.[12] Because of this dispersion of bases it is necessary to divide NATO naval forces into three regions: forces north of the GIUK Gap (i.e. Norway), forces on or around the GIUK Gap (i.e. Britain, Belgium, the Netherlands, West Germany and Denmark) and forces on the American eastern seaboard (i.e. Canada and the USA).

Table 7.2: Soviet Northern Fleet/NATO's Atlantic naval forces by basing areas

	Soviet forces Kola	NATO forces		
		Norway	GIUK Gap	US/ Canada
Ballistic missile submarines	42	0	4	27
Attack submarines	132	14	62	60
Combat aircraft	520	114	434	502
Open ocean warships	42	0	93	113
Coastal warships	95	54	12	0
Mine countermeasures vessels	65	13	33	3
Assault ships (displacement)	38,050	3,900	58,000	340,400
Marines	3,000	0	10,400	45,850
Coastal defence troops	2,200	1,000	0	0

Sources: Simplified from Appendix Table 7A-4.

Dispersing NATO forces to the areas where they are based (see Table 7.2) changes the apparent balance of forces considerably. The simple size comparison (Table 7.1) shows the Soviet Navy to be in a generally weak position. It has more ballistic missile submarines (which would not be utilised in conventional war and in such a war they would be vulnerable and need protecting) and an equal number of attack submarines. However, the Northern Fleet has only half the number of combat aircraft, one-quarter the number of open ocean warships, less than 10 per cent displacement in amphibious assault ships and just 5 per cent of the overall number of available marines, against which the Northern Fleet's slight advantages in coastal warships, mine countermeasures vessels and coastal defence troops are trivial. However, when NATO's forces are dispersed by region the Soviet Northern Fleet's advantage in relation to NATO's forces becomes apparent.

The Northern Fleet could clearly overwhelm the forces stationed in Norway, giving the Soviet Union control of most of the Norwegian and Greenland Seas if NATO's forces at, and south of, the GIUK Gap failed to move northwards. However, the forces in Norway are not without substance. The Norwegian Air Force, though out-numbered and called upon to perform non-naval missions, is modern and powerful. The Norwegian Navy, though probably unable to offer significant resistance to the Northern Fleet in open waters, is a powerful coastal defence force, thus making any Soviet amphibious assault on to the fortified Norwegian coast a potentially very difficult mission.

NATO's forces disposed in and around the GIUK Gap could not prevent Soviet submarines and long-range aircraft penetrating south into the Atlantic through the GIUK Gap.[13] NATO's forces around the GIUK Gap could, however, act as a major restraining force. In particular, submarine and air defence barriers in the GIUK Gap would cause steady, and possibly heavy, attrition of transiting Soviet forces. In order to neutralise fully surface and air threats to the whole region and to ensure the maximum level of attrition against Soviet submarines, NATO's (predominantly US) forces on the eastern coast of America need to be committed. How this might occur depends on where the most pressing needs arise. However, it should not be forgotten that NATO has the advantages of geography in constricting the Northern Fleet. NATO also has superiority in the gross numbers of surface ships, marines and aircraft, and contrary to popular wisdom, equality in the number of attack submarines available — though NATO also has more demanding missions to perform.

Table 7.3: Peacetime deployment of the Soviet Northern Fleet and NATO's northern naval forces

	Soviet forces			NATO forces		
	Kola Area	Norwegian Sea, etc.	Atlantic (& Med)	Norwegian Sea	GIUK Gap	North America
Ballistic missile submarines	30	5	5	5	9	17
Attack submarines	105	12	4 (+11)	14	72	50
Open-ocean warships	39	0	3	0	93	113

Notes: The 'Norwegian Sea, etc.' refers to the Barents, Kara, Greenland and North Seas and the Arctic Ocean. Coastal forces, specialised forces and combat aircraft are assumed to remain largely in the same area as their bases.
Source: Simplified from Appendix Table 7A-5.

DISTRIBUTION BY OPERATING AREAS

The normal peacetime pattern of operations (see Table 7.3) is relatively stable, other than during periodic upsets caused by major naval exercises and the occasional crisis which demands a naval response.

Table 7.4: Presumed initial wartime deployment of the Soviet Northern Fleet and NATO's northern naval forces

	Soviet forces			NATO forces		
	Kola Area	Norwegian Sea, etc.	Atlantic (& Med)	Norwegian Sea	GIUK Gap	North America
SSBNs	14	19	7	8	6	15
Attack subs				34	62	30
Observed (high)	58	52	11 (+11)			
Analysis	71	44	8 (+9)			
Observed (low)	68	42	11 (+11)			
Open-ocean warships	21	18	3	0	117	89
Light frigates	-----45-----		0	7	12	0
FACs	50	0	0	47	0	0
MCM vessels	-----65-----		0	13	33	3
Assault ships	---38,050[a]---		0	--------402,300[a]--------		
Marines	----3,000---		0	---25,400---		30,850

Note: a. Displacement tons
Notes: The 'Norwegian Sea, etc.' refers to the Barents, Kara, Greenland and North Seas and the Arctic Ocean. Combat aircraft have been excluded because their combat radius changes by type of aircraft, type of mission and flight profile. Figures drawn across sea areas denote that this type of force could operate in any of the areas covered, depending upon decisions which are impossible to estimate.
Source: Simplified from Appendix Table 7A-5.

The wartime strategy of NATO's naval commanders will depend on the specific circumstances as they arise and it is quite conceivable that the navies operating in the North Atlantic, Mediterranean and Norwegian Seas could be reinforced in any order as tactically necessary. A NATO initial wartime deployment centred on the North Atlantic and GIUK Gap has been chosen to represent the range of possible deployments (see Table 7.4). From this central position NATO's naval forces could be deployed north or south to support either flank. Such a deployment would also help to seal the barriers at the GIUK Gap, thus ensuring minimum Soviet penetration of the North Atlantic. It would also hold back the considerable striking power of the US 2nd Fleet's four Carrier Battle Groups until Soviet offensive forces in the north or south had had some of their striking power blunted through losses in earlier fighting. Finally it would ensure the minimum disruption to, and maximum protection of, American reinforcements passing across the North Atlantic SLOC, which is SACLANT's highest priority. This initial deployment is no longer favoured by the US Navy in its newly enunciated Maritime Strategy which sees the defeat, or at least containment, of the Northern Fleet as the prime initial task.[14]

Variations in the Soviet wartime deployments largely centre around the degree to which the attack submarine force would be deployed south of the GIUK Gap to interdict the American reinforcements passing along the North Atlantic SLOC. North of the GIUK Gap Soviet attack submarines would perform two very important missions. Firstly, they would provide an Anti-Submarine Warfare (ASW) screen for the whole Northern Fleet and especially for the SSBNs. Secondly, they would be in a position to attack any NATO surface forces, especially the Carrier Battle Groups, that might intrude into the Norwegian Sea.[15] The greater the Soviet commitment to interdicting the SLOC the weaker the Soviet Navy becomes at trying to secure its control over the Norwegian Sea and protecting its SSBNs. The balance of this commitment is a delicate one but behaviour observed during exercises and analysis of the Soviet Navy's mission requirements seem to suggest that most of the attack submarine force will be deployed north of the GIUK Gap rather than against the Atlantic SLOC.[16] In such a deployment the Northern Fleet would be able to maximise its strengths, bringing its submarines, surface ships, and particularly its land-based airpower, into mutual defensive support of a relatively small area. This area could be further protected by the extensive use of minefield barriers, and is in any case granted some immunity by the hostility of the climate. The commonly foul weather above the Arctic Circle offers ice shelters for submarines, and frequent heavy storms to prevent carrier-based air operations and degrade surface ship performance. Combining these factors allows the Soviet Union to create a formidable redoubt around its Kola bases.

AGE: BLOCK OBSOLESCENCE IN THE SOVIET NAVY[17]

Analysis of the age of Soviet and NATO naval forces reveals a serious problem for the Soviet Navy often concealed by aggregate statistics. Analysing the age of the class (i.e. when the first ship or submarine of a new class entered service) rather than analysing the age of individual ships or submarines shows that only 44 per cent of the Soviet Navy vessels are of modern design (over five years to go before reaching the end of the class design's hull life), 6 per cent are old (i.e. within five years of reaching the end of their hull-life) and a full 50 per cent are out of date (over the hull-life of the class design) (see Table 7.5).

Table 7.5: Ship and submarine class age

	Soviet forces			NATO forces		
	Modern	Old	Out of date	Modern	Old	Out of date
	Percentage			Percentage		
Attack submarines:	34	6	60	57	27	16
Nuclear-powered	63	0	37	75	14	11
Diesel/electric powered	13	10	77	41	38	21
Open-ocean warships:	62	8	30	77	13	10
Aircraft carriers	100	0	0	72	14	14
Cruisers	100	0	0	99	1	0
Destroyers/frigates	72	16	12	77	17	6
Gun cruisers/destroyers	23	0	77	0	0	100
Total:	44	6	50	70	18	12
Total less the Diesel/ Electric submarines:	63	4	33	77	13	10

Notes: 'Modern' refers to a class design with over five years to go before reaching the end of the nominal hull-life. 'Old' refers to a class design within five years of reaching its nominal hull-life; whilst 'Out of date' refers to classes which were first introduced beyond the nominal hull-life of the first of the class. This is not a measure of individual ship or submarine age; rather, it is a measure of the age of the class design from when the first of the class entered service.
Source: Abstracted from Appendix Table 7A-6.

The respective figures for NATO are 70 per cent modern, 18 per cent old and 12 per cent out of date. The NATO figures demonstrate a relatively steady building programme of new units replacing the out-of-date vessels, albeit at a moderate rate.[18] The Soviet figures, however, show a lack of continuity which essentially divides the Soviet Navy into two fleets – an out-of-date fleet and a modern one.[19] The very old age of the diesel/electric attack submarine force does distort the analysis for the remainder of the fleet, but even if these are excluded a third of Soviet Navy vessels are out of date. In addition, the figures show a lack of continuity in past procurement with the figure for old vessels much lower than that for the out-of-date ones which have exceeded their anticipated hull-life.

Table 7.6: NATO and Soviet naval construction, 1976–85[20]

	Submarines[21]		Major Surface Combatants[22]	
	NATO	USSR	NATO	USSR
1976	3	10	9	12
1977	3	13	13	12
1978	6	13	19	11
1979	6	12	13	11
1980	3	13	20	11
1981	10	11	17	9
1982	6	8	21	8
1983	7	10	21	10
1984	9	9	16	9
1985	8	7	15	6

Source: US Department of Defense, *Soviet military power* 1981, 1983, 1984 and 1985; Capt. John Moore (ed.) *Jane's fighting ships 1985–86* (Jane's, London, 1985).

Exacerbating this problem is the decline in Soviet naval construction (see Table 7.6) and the relatively long serial production runs which ensure that vessels are still being built when the first vessel of the class is approaching middle age (see Table 7.7).

Table 7.7: The number of vessels built to new designs (and still in service)

Category	Designs 0–10 years old	Designs 11–20 years old	Designs 21–30 years old	Designs over 31 years old
Attack submarines				
NATO	44 (18%)	95 (39%)	88 (36%)	18 (7%)
Warsaw Pact	22 (8%)	76 (26%)	125 (43%)	65 (23%)
Open ocean warships				
NATO	117 (27%)	127 (29%)	144 (33%)	45 (11%)
Warsaw Pact	15 (10%)	59 (38%)	35 (22%)	46 (30%)
Total				
NATO	161 (24%)	222 (33%)	232 (34%)	63 (9%)
Warsaw Pact	37 (8%)	135 (31%)	160 (35%)	111 (25%)

Source: Appendix Table 7A-6.

The age of NATO's fleet is evenly distributed between class designs from less than one to 30 years old, with a few built to designs over 30 years old. In comparison the bulk of vessels in the Soviet Navy are built to designs of between eleven and 30 years old, with only 8 per cent of the fleet built to designs less than ten years old. Thus the Soviet Navy has only 23 per cent of the number of NATO vessels under ten years old in design. The recent introduction of a new series of surface ship and submarine designs, and the probable retirement of some vessels of very old design, will improve this position for the Soviet Navy in the short term despite the decline in shipbuilding, but a long-term improvement will only come about if the Soviet Union reduces the time between the introduction of successive generations. Nonetheless, at present, and for the forseeable future, Soviet classes will lag behind their Western counterparts in their relative age.

The relative slowness with which the Soviet Navy introduces new generations, and the tenacity with which it hangs on to its very oldest generations, exacerbates problems relating to the technical inferiority of Soviet weapons/sensor systems and the comparatively small size of many Soviet warships. The relative technical inferiority of Soviet naval weapons and sensors is widely appreciated and can be demonstrated in a number of ways. Thus, for example, there is the manner in which the West neglected the area of anti-ship missiles in the 1960s, leaving the Soviet Navy a free decade of development in which they ought to have established a commanding lead in this technology. When the West was galvanised into action by the sinking of the Israeli destroyer *Eliat* it overtook the Soviet Union in this area within a little over five years. Another example is a comparison of the Soviet Kirov and US Ticonderoga classes. These vessels offer approximately equal capabilities in air defence and anti-surface warfare, yet to get the same result the Kirov is three times bigger and is being built at a rate of one every three years, against the Ticonderoga's production of three every year; and the Ticonderoga is far superior in anti-submarine warfare. A financial cost comparison between the two is impossible to estimate but the cost in terms of resources is clear: some nine Ticonderogas per Kirov to achieve roughly the same technological performance. This technological inferiority is made worse not only by the relative slowness of introducing successive generations, but by a reluctance to modernise vessels half way through their service life and a reluctance (possibly diminishing) to modernise a class half way through serial production.

Finally, there is the problem of the small size of many Soviet warships, a factor which has distorted many simple static force balances. Partly as a heritage from its coastal navy past and partly as a result of the Soviet preoccupation with defending its frontiers, landward or seaward, the Soviet Navy has an extensive fleet of warships under 1,500 tons. The USA, by way of comparison, has virtually none (and they are mainly Coast Guard vessels). In the main these Soviet vessels are either between 1,000 and 750 tons or between 500 and 200 tons. Those of the latter size are mainly fast-attack craft with essentially coastal duties, while those of larger size lie in the divide between purely coastal vessels and proper open- ocean vessels. Vessels of about 1,000 tons do have a limited ability to sail on the high seas, but this ability is limited by the weather, the ship's endurance and human endurance. Small vessels are susceptible to storm damage and suffer much larger relative movements (such as pitch, roll, yaw and heave) in heavy seas, thus making small vessels difficult to operate as efficient warships in poor weather. Crew performance in particular declines rapidly when they are thrown about the boat by violent wave motions. The ship's endurance, in terms of fuel and weapons payload, is also restricted in a 1,000-ton vessel. For these reasons vessels under 2500 tons have been considered as suitable for operations in coastal sea areas only in this study. This is not to deny their fair-weather capability on the open ocean, but is rather a recognition that fair weather in Northern Waters and the North Atlantic occurs all too infrequently.

Table 7.8: New vessels joining the Northern Fleet as a percentage of the total

	Joining since 1970	Joining since 1980
Ballistic missile submarines	67	100
Attack submarines:		
Cruise-missile armed	65	100
Torpedo-only armed	61	100
Open ocean warships:-		
Aircraft carriers, cruisers		
and gun cruisers	40	64
Destroyers and gun destroyers	30	75
3,900-ton frigates	25	...

Sources: The Norwegian Atlantic Committee, *Excerpt from the Norwegian version of The Military Balance 1984–85* (Norwegian Atlantic Committee, Oslo, 1985). Tomas Ries, 'Defending the far north', *International Defense Review*, no. 7 (1984), p. 874.

Some of the decline in shipbuilding noted in Table 7.6 is a result of a declining Soviet interest in warships of this small size. Even the maritime border guard section of the KGB, whose largest ships used to be of the 1,000-ton type, has begun to take delivery of much larger ships, the Krivak III, derived from the Navy's 3,700-ton general-purpose frigate.

Despite this largely gloomy picture for the Soviet Navy, the Northern Fleet has prospered, taking the lion's share of new construction since 1970 and almost all new vessels since 1980 (see Table 7.8). This has coincided with the build-up of the Northern Fleet's SSBN force, and consequently the size and importance of the Northern Fleet is almost certainly directly linked with the future of the SSBNs.

CONCLUSION

The exact nature of the naval force balance during wartime in Northern Waters depends largely upon the prevailing circumstances at the start of a conflict. If the genesis of an East–West conflict were outside the NATO area it is possible that the initial naval deployments, particularly of the US Navy, would be inappropriate, placing SACLANT in a situation where his ability to operate in the Norwegian Sea against determined Soviet opposition would be in jeopardy. Likewise, if the defence of the North Atlantic SLOC requires enhanced capability because of greater Soviet efforts in SLOC interdiction, then escort forces for the carriers could become scarce, or the carriers themselves might be dedicated, directly or indirectly, to supporting the SLOC. Thus, it is quite conceivable that NATO's initial wartime deployment could be less than optimal for supporting the Northern Flank — though the degree to which this is a serious risk depends on when the political decision is taken to alter peacetime naval diplomacy deployments into wartime dispositions.

The Soviet High Command must in turn estimate the risk involved in pursuing an aggressive anti-SLOC campaign, thereby leaving his defences in the Norwegian Sea weaker but reducing the chances of Carrier Battle Groups operating there, or holding back the bulk of his forces and awaiting any offensive moves by NATO. The basis of these calculations must be the accomplishment of two vital missions in Northern Waters: the NATO commander must keep the SLOC open and ensure the defence of northern Norway; and the Soviet commander must maintain the integrity of the SSBN force and ensure that the Soviet Union does not come under direct attack from this northern seaward flank. Both of these tasks demand the control of the sections of Norwegian Sea.

At current force levels, if SACLANT is allowed by circumstances to concentrate on supporting the NATO's Northern Flank, it seems most probable that NATO will prevail in the southern and central Norwegian Sea, despite the investment of the Soviet Navy's most modern vessels in this area. If circumstances deny SACLANT the necessary resources, then it seems likely that the Soviet Northern Fleet will retain control of the whole Norwegian Sea, not merely the northern-most segment under land-based fighter cover. However, the Norwegian Sea has no natural geographic divide so that any apportioning of that sea — Soviets in the north and NATO in the south — will lead to intense engagements at the notional boundary as both sides try to secure their respective areas of interest from hostile interference. The rough position of such an engagement area will probably be determined by the limit of effective land-based fighter cover from Soviet and NATO bases. Submarine warfare, however, with no natural constricting choke point between the North Cape and the GIUK Gap, will have a free range for combat throughout the length of the Norwegian Sea. The extent to which either side can achieve sea control over significant portions of the Norwegian Sea will depend upon how close either side approaches the other's base areas; the further north NATO naval forces proceed the less likely their chances of achieving effective sea control against mounting opposition and vice versa for Soviet deployments southwards.

Predicting possible outcomes for a large-scale naval war in the Norwegian Sea is beyond the scope of this chapter, which has merely tried to set out the starting positions of the two sides in such a conflict. Clearly, each side faces trade-offs and risks in how aggressively or defensively it conducts its respective campaigns. However, barring unforeseen political developments which could wrongfoot NATO's initial wartime deployment, it is evident that NATO has the flexibility to capitalise on substantial superiorities in SSBN invulnerability, naval air warfare, surface warfare and the power projection capabilities of its marines. The Northern Fleet holds the quantitative edge only in mine warfare and in limited range coastal vessels. The highly publicised Soviet submarine threat is in fact equalled, quantitatively, and bettered, qualitatively, by NATO. In addition, given the relative weakness of Soviet surface and naval air forces and the corresponding vulnerability of Soviet SSBNs, the offensive role of Soviet attack submarines will probably be restrained by the need for these submarines to supplement the active defence of the SSBNs. NATO's attack submarines, beyond manning the ASW barrier at the GIUK Gap and providing escorts for aircraft carriers, will not have their offensive qualities diminished by defensive constraints. When all this is coupled with Soviet limitations due to the advanced age, small size, and lack of technological sophistication of many of their naval units, it is clear that NATO, on balance, would begin a naval conflict in the Atlantic and Northern Waters in a far healthier condition than it is normally given credit for, and this is a condition NATO must continue to maintain.

APPENDIX

Table 7A-1: The composition of the Red Banner Northern Fleet, 1975–85

	1985	1980	1975
STRATEGIC NUCLEAR FORCES			
Nuclear-powered ballistic missile submarines	40	48	38
THEATRE NUCLEAR FORCES			
Diesel/electric ballistic missile submarines	2	8	15
GENERAL PURPOSE NAVAL FORCES (Open-ocean)			
Attack submarines	132	90	119
Combat aircraft	380	302	?
Open-ocean ships	42	33	32
Attack submarine details:			
(Cruise missile - nuclear-powered)	30	28	28
(Cruise missile - diesel/electric)	7	8	16
(Torpedo - nuclear-powered)	45	34	24
(Torpedo - diesel/electric)	50	20	51
Combat aircraft details:			
(Fighters)	0	0	?
(Strike fighters)	12	12	?
(Strike bombers)	100	80	?
(Anti-submarine aircraft)	50	30	?
(Anti-submarine helicopters)	133	95	?
(Tankers, reconnaissance and EW aircraft)	85	85	?
Open-ocean ship details:			
(37,000-ton Kiev-class aircraft carriers)	1	1	0
(28,000-ton Kirov-class cruisers)	2	1	0
(7,500-12,500-ton cruisers)	9	9	7
(3,500-8,000-ton destroyers)	17	9	8
(3,900-ton frigates)	8	6	1
(17,200-ton gun cruisers)	2	2	3
(3,130-3,500-ton gun destroyers)	3	5	13
GENERAL PURPOSE NAVAL FORCES (coastal and specialised forces)			
Coastal ships	95	132	86
Mine countermeasures vessels	65	66	90
Assault ships (displacement tonnage)	38,050	28,800	17,600
63rd Naval Infantry Brigade	3,000	1,800	1,800
Coastal Artillery and Rocket Troops Regiment	2,200	2,200	2,200
Coastal ships details:			
(1,150-1,500-ton light frigates)	45	46	28
(210-660-ton missile craft and corvettes)	25	41	28
(170-580-ton torpedo and fast patrol craft)	25	45	30

Notes: All displacement figures are in Imperial long tons at full load. The low figure for attack submarines in 1980, caused by the low number of diesel–electric torpedo submarines, is against the general trend and if the estimate of 50 is used the total becomes a more consistent 120. Likewise, the figures for coastal warships in 1980 are unusually high.

Sources: Captain John Moore, RN (ed.), *Jane's fighting ships 1975–76*, (Jane's, London) p. 549, 1980–81 p. 465, 1985–86, p. 510; Breyer and Polmar *Guide to the Soviet Navy, 2nd edition* (United States Naval Institute, Annapolis, 1977), pp. 501–3; International Institute for Strategic Studies (IISS) *Military Balance 1984–85*, (IISS, London, 1984), p. 21; Tomas Ries, 'Defending the far north', *International Defense Review*, July 1984, p. 878.

Table 7A-2: The composition of forces associated with the Kola Peninsula (Leningrad Military District (LMD) and Arkhangelsk Voyska PVO District)

NORTH-WESTERN TVD (HQ Petrozavodsk)

	On Kola	Outside Kola	Total

AIR DEFENCE FORCES

	On Kola	Outside Kola	Total
Voyska PVO defence aircraft (Interceptors, interceptors/fighter, FGA — MiG-23, MiG-25, MiG-29, MiG-31, Su-15, Yak-28)	120	214	334
(Airborne Early Warning — TU-126 Moss)	0	6	6

Voyska PVO air defence missiles:
  on Kola 180 launchers in 30 SAM complexes
  outside Kola 120 launchers in 20 SAM complexes
6th Army SAM Brigade: 60 launchers in 3 regiments

AIR FORCES

	On Kola	Outside Kola	Total
LMD Regional Air Force Command (HQ Leningrad)	130	370	500
Details: (FGA — MiG-21, MiG-27, Su-17)	0	130	130
(Reconnaissance — MiG-21/25, Su-17)	30	30	60
(Attack helicopters)	30	60	90
(Utility helicopters — Mi-8/6/2)	55	120	175
(Utility aircraft)	15	30	45

Other air forces:
  Baltic Fleet naval aviation — 40 Backfires using Olenegorsk Air Base

GROUND FORCES

Leningrad Military District (HQ Leningrad)
6th Army (HQ Petrozavodsk)

	On Kola	Outside Kola	Total
27th Corps (HQ Arkhangelsk) and 30th Corps (HQ Vyborg)			
Motor rifle divisions	2	7	9
Airborne divisions	0	1	1
Artillery divisions	0	1	1
Artillery brigade	1	?	1-?
Rocket artillery brigade	1	?	1-?
SAM brigade	1	?	1-?
Air-assault regiments	1-2	1-2	3
Other forces: 63rd Naval Infantry Brigade	1	0	1

FORCES UNDER OUTSIDE CONTROL

Naval SPETSNAZ Brigade (Main Intelligence Directorate — GRU)
6 SS-5 Skean silos (strategic rocket forces)
Hen House perimeter acquisition radars (Voyska PVO Moscow Galosh ABM)
Advance basing for 36th Air Force (strategic bombers and tankers)
Advance basing for 46th Air Force (theatre bombers)

Sources: John Berg, 'Soviet Front-level threat to northern Norway', *Jane's Defence Weekly*, 2 February 1985, pp. 178–9; Tomas Ries, 'Defending the far north' *International Defense Review*, July 1984, pp. 875–8; US Department of Defense, *Soviet military power, 4th Edition*, (US Government Printing Office, Washington DC, 1985), pp. 13, 33, 104, 107; IISS, *The military balance 1985–86*, (IISS, London, 1985), p. 19.

Table 7A-3: NATO naval forces available for operations in the Norwegian Sea, North Sea (excluding the English Channel), Barents Sea and North Atlantic

	Norwegian Sea area	GIUK Gap area	Canada	US 2nd Fleet	Total
STRATEGIC NUCLEAR FORCES					
Ballistic missile submarines	0	4	0	27	31
GENERAL PURPOSE NAVAL FORCES (Open-ocean)					
Attack submarines	14	62	3	57	136
Combat aircraft	114	434	68	775	1,391
Open-ocean ships	0	93	16	97	206
Attack submarines details:					
(Cruise-nuclear-powered)	0	0	0	3	3
(Torpedo-nuclear-powered)	0	14	0	53	67
(Torpedo-diesel/electric)	14	48	3	1	66
Combat aircraft details:					
(Fighter aircraft)	0	96	0	72	168
(Strike fighters)	107	84	0	48	239
(Strike bombers)	0	25	0	108	133
(Maritime patrol aircraft)	7	49	33	114	203
(ASW helicopters)	0	143	35	59	237
(AEW, EW, tanker aircraft)	0	37	0	63	100
Open-ocean ship details:					
(Aircraft carriers)	0	0	0	4	4
(Assault carriers)	0	0	0	5	5
(ASW carriers)	0	3	0	0	3
(Battleship)	0	0	0	1	1
(10,500–8,200-ton cruisers)	0	0	0	10	10
(8,300–4,000-ton destroyers)	0	30	4	31	65
(4,000–2,500-ton frigates)	0	60	12	46	118
GENERAL PURPOSE NAVAL FORCES (coastal and specialised forces)					
Coastal ships	54	12	0	0	66
Mine countermeasures vessels	13	33	0	3	49
Assault ships (displacement)	3,900	58,000	0	228,700	290,600
Marines	0	10,400	0	45,850	56,250
Coastal defence troops	1,000	0	0	0	1,000
Coastal ships details:					
(2,500–1,000-ton frigates)	5	7	0	0	12
(–1,000-ton corvettes)	2	5	0	0	7
(Missile fast-attack craft)	39	0	0	0	39
(Torpedo and fast-patrol craft)	8	0	0	0	8

Notes: The following forces have been excluded: *All air forces* (Denmark, Belgium and West Germany). *Some air forces* (Britain, Holland, Canada and USA). *All Pacific and Mediterranean forces* (US and Canada). All coastal forces (Denmark and West Germany). *Mine countermeasures vessels less than 500 tons* (Britain, Holland and Belgium). *All forces* (France, Spain and Portugal). US 6th Fleet assumed to contain two Carrier Battle Groups, one Amphibious Squadron and one Underway Replenishment Group, plus 4 Ballistic Missile Submarines. The Assault Carriers add 111,700 tons to the Assault Ships displacement tonnage, for 340,400 tons (USA) and 402,300 tons (total).

Sources: IISS, *Military balance 1984–85*, (IISS, London, 1984); Jean Labayle Couhat and A.D. Baker III (eds), *Combat fleets of the world 1984/85*, (Arms and Armour Press, London, 1984; Captain John Moore RN (ed.), *Jane's fighting ships 1985–86* (Jane's Publishing Company, London, 1985); Norman Polmar, *The ships and aircraft of the US Fleet 13th Edition* (Arms and Armour Press, London, 1985); Congressional Budget Office, *The US Sea Control Mission: forces, capabilities and requirements* (US Government Printing Office, Washington, DC, 1977), pp. 10–13, 31, 54–5.

Table 7A-4: Soviet Northern Fleet/NATO's Atlantic naval forces by basing areas

	Soviet forces Kola	NATO forces		
		Norway	GIUK Gap	US/ Canada
Ballistic missile submarines:				
Nuclear-powered	40	0	4	27
Diesel/electric-powered	2	0	0	0
Attack submarines:	132	14	62	60
Nuclear-powered: cruise missile	30	0	0	3
torpedo	45	0	14	53
Diesel/electric: cruise missile	7	0	0	0
torpedo	50	14	48	4
Combat aircraft:	520	114	434	502
Fighter aircraft	120	0	96	72
Strike fighters	12	107	84	48
Strike bombers	100	0	25	108
ASW aircraft	50	7	49	117
ASW helicopters	133	0	143	94
EW, AEW, tanker and reconnaissance	105	0	37	63
Open-ocean warships:	42	0	93	113
Aircraft carriers	0	0	0	4
Assault carriers/Kiev	1	0	0	5
ASW carriers	0	0	3	0
Battleships/Kirov	2	0	0	1
Cruisers	9	0	0	10
Destroyers	17	0	30	35
Frigates	8	0	60	58
Gun cruisers	2	0	0	0
Gun destroyers	3	0	0	0
Coastal warships:	95	54	12	0
Light frigates and corvettes	45	7	12	0
Fast-attack craft	50	47	0	0
Mine countermeasures vessels:	65	13	33	3
Assault ships (displacement):	38,050	3,900	58,000	228,700
Marines:	3,000	0	10,400	45,850
Coastal defence troops:	2,200	1,000	0	0

Note: Elements of one Motor Rifle Division (45th) also have amphibious training. Combat Aircraft in Kola include units of the Leningrad Military District and the Arkhangelsk Voyska PVO District.
Sources: Abstracted from Appendix Tables 7A-1, 7A-2 and 7A-3.

Table 7A-5: Soviet Northern Fleet/NATO's Atlantic naval forces by presumed initial wartime and peacetime operating areas

BALLISTIC MISSILE SUBMARINE DEPLOYMENTS

	Soviet forces			NATO forces		
	Kola Area	Norwegian Sea, etc.	Atlantic	Norwegian Sea	GIUK Gap	North America
Wartime	14	19	7	8	6	15
Peacetime	30	5	5	5	9	17

Note: 'Norwegian Sea, etc.' refers to the Barents, Kara and Greenland Seas and on occasion possibly also the North Sea and Arctic Ocean.

Soviet peacetime SSBN deployments are on average three Yankee class[23] and, in response to the deployment of cruise and Pershing missiles, two Delta class, in the North Atlantic.[24] About the same number would be in transit to and from the deployment area.[25] The wartime figures assume a surge deployment of 50 per cent of all active units (less those already deployed).[26] Seven of the 19 assigned to the 'Norwegian Sea' in wartime are in transit to and from the North Atlantic, while one is a Golf class SSB assigned to theatre nuclear missions, leaving eleven Deltas and Typhoon class SSBNs on station for strategic nuclear missions. It is possible that some of the 21 Yankee class have been assigned to theatre missions.[27]

NATO's peacetime SSBN deployments comprise the following. One UK Polaris SSBN on patrol and three in the UK (GIUK Gap area). Twelve US Trident I submarines either in the USA or deployed in the North Atlantic. The remaining 19 US Poseidon submarines forward deployed. Here we assume that the split is near even, with ten deployed northwards (via Holy Loch) and nine southwards (via Rota) into the Mediterranean. These both support four US SSBNs on station in the 'Norwegian Sea' and in the Mediterranean (the four SSBNs actually assumed to be in the Mediterranean are excluded from this and other tables). The 'Norwegian Sea' area only implies one of the more likely patrol areas — details of US and UK SSBN patrol areas are classified. NATO's SSBNs operate at a very high tempo in peacetime leaving few units which have to, or could be further deployed. It is assumed here that one UK and two US submarines leave the GIUK Gap area for the 'Norwegian Sea'. A further two Poseidon SSBNs supporting the Mediterranean deployment leave the USA.[28]

ATTACK SUBMARINE DEPLOYMENTS

	Soviet forces			NATO forces		
	Kola Area	Norwegian Sea, etc.	Atlantic & Med.	Norwegian Sea	GIUK Gap	North America
Wartime				34	62	30
Observed (high)	58	52	22			
Analysis	71	44	17			
Observed (low)	68	42	22			
Peacetime	105	12	15	14	72	50

The peacetime distribution of Soviet Northern Fleet attack submarines is taken from James John Tritten.[29] Eleven attack submarines are assumed to be in the Mediterranean, with three in the North Atlantic and one in the Caribbean (though this latter is not a regular patrol). Those attack submarines in the 'Norwegian Sea' are submarines in transit from the Mediterranean/North Atlantic and could be anywhere along those routes.

The wartime allocation is based on the deployments during naval exercises and the work of Paul H. Nitze et al.[30] Major Northern Fleet exercises in April 1984 and July 1985 deployed 20 and 32 Northern Fleet attack submarines (low and high observed assessments) respectively into the Norwegian Sea[31] — the additional submarines in the Norwegian Sea are in transit. The Nitze et al. study analyses the allocation of attack submarines to different missions. By using the same ratios found in the study results and estimating the geographic location where these missions are likely to be mainly carried out, a wartime distribution can be drawn up. This analysis closely follows observed exercises behaviour. The analysed missions, allocation and main operating areas are:

- Command of Contiguous Waters, 21 Submarines in Norwegian Sea, etc.
- SSBN Protection, Outer Area Attack, Counter Carrier and Counter SSBN, 60 Submarines in Norwegian Sea, etc.
- Port and Sea Lines of Communication Interdiction, 25 Submarines in North Atlantic.
- Support Mediterranean Squadron, 26 Submarines in Mediterranean.

For the table above these allocations have been reduced by two-thirds (one-third in port for various reasons and one-third in transit). The one-third in transit to or from the North Atlantic/Mediterranean has been added to the submarines in the 'Norwegian Sea' area.

NATO attack submarine peacetime deployments are assumed to be essentially similar to the basing deployment (see Appendix Table 7A-3) other than the deployment of some US submarines to the GIUK Gap. The wartime deployments are very rough estimates. Twenty (one-third) of GIUK Gap-based submarines are assumed to go northwards into the Norwegian Sea. Twenty (one-third) US submarines are assumed committed to the GIUK Gap. Ten US submarines are sent to reinforce the Mediterranean. All these figures are speculative and are included only for illustrative purposes.[32]

SURFACE COMBATANT DEPLOYMENTS: OPEN-OCEAN WARSHIPS

	Soviet forces			NATO forces		
	Kola Area	Norwegian Sea	Atlantic	Norwegian Sea	GIUK Gap	North America
Wartime	21	18	3	0	117	89
Peacetime	39	0	3	0	93	113

Soviet peacetime open-ocean warship deployments to the Atlantic include stationing at the West Africa patrol and occasional visits to the Caribbean.[33] The Soviet wartime deployment is based on the July 1985 exercise.[34]

The allocation of NATO's warships in wartime is very variable since naval forces may be in demand from four possible areas: into the Norwegian Sea for the reinforcement of Norway; into the North Atlantic to secure the SLOCs;

into the Mediterranean for the reinforcement of the 6th Fleet and out of the NATO area for some other contingency. The figures above assume that the US 2nd Fleet's four Carrier Battle Groups in the region of (but south of) the GIUK Gap. Since venturing north of the GIUK could entail the loss of a national asset (i.e. a carrier) it is unlikely to occur unless there is some specific operation in mind, such as interfering with major Soviet naval operations or undertaking a marine landing in Norway.[35] Peacetime deployments are the same as the basing areas but NATO regularly exercises in all sea areas (e.g. exercise TEAMWORK for the reinforcement of Norway and exercise OCEAN SAFARI for securing the SLOC).

SURFACE COMBATANT DEPLOYMENTS: COASTAL AND SPECIALISED FORCES

	Soviet forces			NATO forces		
	Kola Area	Norwegian Sea	Atlantic	Norwegian Sea	GIUK Gap	North America
Light frigates Wartime	-----45-----		0	7	12	0
Fast-attack craft Wartime	50	0	0	47	0	0
Mine counter-measures vessels	-----65-----		0	13	33	3
Assault ships	38,050 tons	0		-----402,300 tons-----		
Marines	----3,000----	0		---25,400---		30,850

Soviet forces which are participating in amphibious operations or other tasks in support of a land campaign in northern Norway will come under the operational command of the Army Front controlling these land advances into Norway. This front, variously known as the Arctic Front or Northern Front, will be formed from the headquarters of the Leningrad Military District. This could include gun-armed cruisers as well as coastal forces. Purely naval tasks, such as minesweeping, will generally remain under naval control unless needed to support a specific mission in support of the land campaign.[36] Some coastal forces operated by the KGB's maritime border guards are likely to come under naval control during war.

The full amphibious lift of the US 2nd Fleet can transport one US Marine Brigade. Such a Brigade (15,000 men) committed to reinforce Norway will need much less sealift because its heavy equipment is being pre-positioned in central Norway. However, this pre-positioning programme will not be complete until 1989 and it will of course demand resources from strategic air lift. Coastal forces and mine countermeasures forces are generally limited by range from operating too far from the basing areas.

Table 7A-6: The age of submarine and ship classes design: The age of warship and submarine class design, as of 1/1/86, by hull-life from the date of the commissioning of the first unit of a class

CATEGORY	0–10	11–20	21–25	26–30	31–35	36–40	41–45	Total Number	Old	Out-of-date
Warsaw Pact navies										
SSGN	2	19	..	[28	..	..	..]	49	..	28
SSG	..	..	16	[..	..	..	..]	16	16	..
SSN	16	39	..	[17	..	..	..]	72	..	17
SS	4	18	..	[64	4	61	..]	151	..	129
Submarines	22	76	16	109	4	61	..	288	16	174
CVS	..	6	..	..	..	[..	..]	6	..	..
CG	15	21	..	..	[..	..	..]	36	..	..
DD	..	32	23	12	[9	..	..]	76	12	9
Gun ships	..	..	..	..	[28	9	..]	37	..	37
Oceanic ships	15	59	23	12	37	9	..	155	12	46
Sub. & ship	37	135	39	121	41	70	..	443	28	220
NATO navies										
SSGN	34	50	13	[..	..	..	..]	97	13	..
SSN	..	..	3	[12	..	..	..]	15	3	12
SS	10	45	50	[10	4	..	14]	133	50	28
Submarines	44	95	66	22	4	..	14	245	66	40
CV/CVS	4	3	8	..	3	[..	3]	21	3	3
CG	11	36	22	1	[..	..	..]	70	1	..
DD	102	87	54	54	[..	2	17]	316	54	19
Gun ships	..	1	5	..	[..	..	20]	26	..	20
Oceanic ships	117	127	89	55	3	2	40	433	58	42
Sub. & ship	161	222	155	77	7	2	54	678	124	82

Glossary: SSGN (Nuclear-powered Cruise Missile Armed Submarine), SSG (Diesel/Electric-powered Cruise Missile Armed Submarine), SSN (Nuclear-powered Torpedo Armed Submarine), SS (Diesel/Electric-powered Torpedo Armed Submarine), CV (Aircraft Carrier), CVS (Anti-submarine Aircraft Carrier), CG (Cruiser), DD (Destroyer), Gun Ships (Gun-armed Battleships, Cruisers and Destroyers).

Notes: This table shows the age of ship classes, not of individual ships themselves. Thus, it is the design age and not the ship age which is being displayed. Only designs thought to be in service by 1/1/85 are included and the age of a class design is dated from 1/1/86. The brackets [ ] denote where the designs in a particular category are thought to be 'out of date' based on estimated hull-life of vessels. Vessels within five years of being out of date (i.e. those vessels in the left column from the first '[') are considered as 'old'.

The hull-lives are: 35 years for Carriers; 30 years for Cruisers, Destroyers and Gun Ships; 25 years for Submarines.[37] Modernisations, such as Carrier SLEP or Battleship reactivation programmes, are not generally taken into account.
Source: Jean Labayle Couhat and A.D. Baker III (eds), *Combat fleets of the world 1984/85: their ships, aircraft, and armaments* (Arms and Armour Press, London, 1984).

NOTES

1. International Institute for Strategic Studies (hereafter IISS), *Military balance 1984–85*, (IISS, London, 1984), p. 21; IISS, *Military balance 1985–86*, (IISS, London, 1985), p. 26.
2. Tomas Ries, 'Defending the far north', *International Defense Review* (hereafter IDR), no. 7, (1984), p. 875; IISS, *Military balance 1985–86*, (IISS, London, 1985), p. 26.
3. Including the Kara, Barents, Greenland and Norwegian Seas and the Arctic Ocean; and southwards through the GIUK Gap into the North Sea and North Atlantic. The Northern Fleet also provides vessels for more distant warm-water areas. The most important of these is the Mediterranean, where the Northern Fleet is responsible with the Baltic Fleet for supplying submarines. Vessels are also periodically supplied to the Caribbean and South Atlantic.
4. The distribution will be analysed by basing area, by peacetime operating area and by a presumed initial wartime operating area.
5. A hypothetical war between NATO and the Warsaw Pact could, for example, start on the Southern Flank, in the Mediterranean, perhaps growing out of a Middle East conflict, thus possibly draining away forces that could have been used further north during the initial crisis period and the start of a conflict. The Northern Fleet is responsible for the supply of the bulk of Soviet submarines in the Mediterranean, while the US 2nd Fleet in the Atlantic could be weakened to reinforce the US 6th Fleet in the Mediterranean. Once a conflict had started in earnest it is likely that units would be withdrawn from the Mediterranean back into the strategically more important Atlantic, but initially this could lead to NATO's naval forces being temporarily wrongfooted.
6. The definitions of ship/submarine types are as follows: 'ballistic missile submarines' includes SSBNs and diesel–electric powered ballistic missile carrying submarines. 'Attack submarines' includes all others, whether nuclear or diesel–electric powered, or whether armed with cruise missiles or only torpedoes. Open-ocean warships are all those general purpose armed naval vessels over 2,500 tons full load displacement, coastal warships being all those under this figure. 'Mine countermeasures vessels' includes both minesweepers and minehunters. The assault ships are the amphibious lift for the marine forces and are measured by tonnage (full load displacement) rather than by number. Full details of forces included and excluded in this table can be found in Appendix Tables 7A-1, 7A-2 and 7A-3.
7. As an example, the Naval Aviation of the Baltic Fleet contains a force of some 40 Backfire bombers. These have been used in support of Northern

Fleet exercises, using air bases on the Kola Peninsula for advanced basing. See Hugh Lucas and Antony Preston, 'USSR's Navy exercise' *Jane's Defence Weekly* (hereafter JDW), 14 April 1984, p. 547.

8. Units of the Baltic and Black Sea Fleets regularly participate in North Atlantic/Norwegian Sea exercises which are predominantly Northern Fleet affairs. However, neither of these fleets is factored into the Soviet calculations presented here. This is because neither of them is likely to force a passage through their respective choke points until quite late in a war, if at all. It is therefore unlikely that the Northern Fleet relies on such outside help to achieve its missions. Such Baltic or Black Sea Fleet units deployed beyond their choke points at the beginning of a conflict would probably be placed under Northern Fleet control since such units would only be capable of returning to Soviet bases in the Kola area. The units are unlikely to be caught in significant numbers. The participation of these two fleets in Northern Fleet exercises is therefore likely to be for other reasons, such as ensuring a homogeneous level of training throughout the Navy, or supplying Northern Fleet admirals with adversaries not personally known to them. NATO naval forces deployed in the Baltic and Mediterranean Seas have likewise been excluded, although NATO could redeploy these units to the Atlantic with much greater ease since they would not have to force any choke points during their transit.

9. The GIUK Gap runs between Greenland and Iceland (the Denmark Strait) and between Iceland and the United Kingdom (with the Faroes in the middle). It could also be said to include the northern entrance to the North Sea between the UK and Norway and the English Channel.

10. For fuller details of NATO forces included and excluded in this assessment, see Appendix Table 7A-3.

11. The main bases are Servomorsk (the Fleet headquarters) and Polyarnyi (the main submarine base) down river from Murmansk (a largely commercial harbour) on the Kola fjord. Other bases are Lista, Ura Guba, Sayda, Olenya and, further along the Kola Peninsula coast than the rest, Iokanga (base for Typhoon-class submarines). See William M. Arkin and Richard W. Fieldhouse, *Nuclear battlefields: global links in the arms race* (Ballinger, Cambridge, Massachusetts, 1985), Appendix B, pp. 252–63; Tomas Ries, 'Defending the far north', p. 873; US Department of Defense, *Soviet military power 1985*, 4th edn (US Government Printing Office, Washington, DC, 1985), p. 26.

12. NATO's main naval bases are: in the Atlantic: Norfolk (headquarters of US and NATO naval forces in the region), Mayport, Charleston, Jacksonville, Brunswick, New London, New York, Boston, New Orleans, Bangor, King's Bay (all US); Holy Loch (in UK); Keflavik (in Iceland); Bermuda. In the North Sea: Kallo, Ostend, Zeebrugge (Belgium); Devonport, Faslane, Portland, Portsmouth, Rosyth (United Kingdom); Halifax (Canada); Frederikshavn (Denmark); Wilhelmshaven (West Germany); Den Helder, Flushing (Netherlands). In the Norwegian Sea: Horten, Bergen, Ramsund and Tromsø (Norway). Source: IISS, *Military balance 1985–86*, (IISS, London, 1985), *passim*.

13. From bases in the Kola Peninsula only Bear reconnaissance, electronic warfare, and anti-submarine aircraft and Backfire bombers would be able to travel the 1,400 plus nautical miles to the North Atlantic through the GIUK Gap. Badger bombers could reach the area of the GIUK Gap. However, fighters

such as the Fencer or Flogger can only cover the northern half of the Norwegian Sea at the limit of their operational radii. The capture of north Norwegian air bases would shorten the range needed and so increase the air threat. Likewise, the capture of air bases in West Germany or Denmark could increase the air threat, especially if the UK air defence network were severely degraded. For a good, though a little dated, discussion of the air defence of the GIUK Gap, see Congressional Budget Office (hereafter CBO), *The US Sea Control Mission: forces capabilities and requirements*, (US Government Printing Office, Washington DC, 1977), Chapters 3, 4 and Appendix A.

14. For a long time the US Navy evaded the issue of whether it planned to concentrate, in the NATO area, on the Norwegian Sea. See, for example, the protracted discussion between Senator Sam Nunn and Secretary of the Navy John Lehman in *Hearings before the US Senate Subcommittee on Sea Power and Force Projection on Department of Defense Authorization for Appropriations for Fiscal Year 1985, Wednesday, March 14, 1984*, pp. 3871–3, pp. 3878–9, p. 3893 and *passim*. See also the correspondence between Larry Bond and Tomas Ries for a discussion of whether the 2nd Fleet Carrier Battle Groups would be sent north of the GIUK Gap or remain in the North Atlantic, in 'Controversy: a new strategy for the North-east Atlantic?', *IDR*, no. 12 (1984), pp. 1803–4. Very recently, however, this strategy has been formally acknowledged, see Michael R. Gordon, 'In nonnuclear war, US Navy might hit Soviet subs', *International Herald Tribune*, 8 January 1986, p. 1; Admiral Wesley McDonald (SACLANT) 'Mine warfare: a pillar of maritime strategy', *US Naval Institute Proceedings*, October 1985; Major Hugh K. O'Donnell, Jr., 'Northern Flank maritime offensive', *US Naval Institute Proceedings*, September 1985; and especially, Admiral James D. Watkins (CNO), 'The Maritime Strategy', *US Naval Institute Proceedings*, January 1986.

15. Keeping NATO Carrier Battle Groups out of the Norwegian Sea would also indirectly help to protect the Soviet SSBNs since their presence would make the submarine bases vulnerable to air attack, allow additional NATO air and surface ASW units into the area to attack the Soviet SSBNs and their escorts, and divert Soviet units away from screening the SSBNs. For a useful brief survey of development of Soviet naval missions, see CBO, *Shaping the general purpose navy of the eighties: issues for fiscal years 1981–1985* (US Government Printing Office, Washington, DC, 1980), Chapter 1, especially pp. 23–33.

16. Soviet Northern Fleet exercises in 1983, 1984 and 1985 were reported to have large-scale submarine deployments, but these were all either in, or north of, the GIUK Gap. See Desmond Wettern, 'Stretching the bear's claws: Soviet exercise '83', *Navy International*, September 1983, pp. 518–24; Air Commodore G.S. Cooper, 'Soviet naval exercise stretches NATO forces', *Daily Telegraph* (London), 19 July 1985, p. 4. See also the study of the Atlantic Council Working Group on Securing the Seas, in which their analysis (in Chapter 4) suggested that the Soviet Navy might allocate (not deploy) 65 submarines to the anti-SLOC mission world-wide (half of which would be in the North Atlantic). This allocation would roughly equal the deployments assumed here, with about nine submarines deployed south of the GIUK Gap on the anti-SLOC mission. The effect of such an allocation is analysed in an interesting fashion in Chapter 13. Paul H. Nitze, Leonard Sullivan, Jr and the

Atlantic Council Working Group on Securing the Seas, *Securing the seas: the Soviet naval challenge and Western Alliance options*, (Westview Press, Boulder, 1979), pp. 108–16, 349–60. For the threat of mine warfare in this region, see Tom Stefanick, *Starfish wars: strategic anti-submarine warfare and naval strategy* (Institute for Defence and Disarmament, 1986), cited in 'Soviets sowing mines', *Navy News and Undersea Technology*, 6 December 1985, p. 8.

17. This section is derived from a study by the author on the NATO–Warsaw Pact naval force balance, to be published soon as a *Whitehall Paper* by the Royal United Services Institute for Defence Studies (RUSI).

18. In fact NATO's old and out-of-date vessels are concentrated almost wholly in a few navies that have had long-term under-funding or have had military assistance — mainly from the USA — withdrawn because of political circumstances. These countries are Canada, Turkey, Greece, Spain and Portugal. Canada and Spain are undertaking urgent measures to redress the neglect of their navies. Portugal is seeking NATO funding to improve its navy, whilst Greece and Turkey have been improving certain aspects of their navies, particularly their coastal forces.

19. The Soviet figures actually include the whole Warsaw Pact, but the non-Soviet Warsaw Pact navies are so small their inclusion or exclusion does not materially affect the figures.

20. The NATO figures exclude French and Spanish naval construction. References for US Department of Defense *Soviet military power* are 1981, p. 13; 1983, p. 79; 1984, p. 99; 1985, p. 105.

21. Submarines include attack submarines and ballistic missile submarines.

22. 'Major surface combatants' includes all general-purpose warships over 1,000 tons.

23. Commander Richard T. Ackley, USN (ret.), 'The wartime role of Soviet SSBNs', *US Naval Institute Proceedings*, vol. 104, no.6 (June 1978), p. 36.

24. RUSI, *News sheet 37: February 1984* (RUSI, London, 1984), p. 1.

25. James John Tritten, *Soviet Navy data base: 1982–83* (The Rand Corporation, Santa Monica, 1983), p. 75, fn. 30.

26. This assumption comes from ibid., p. 39, where it describes a high rapid surge threat. Other scenarios featured a low surge (33.3 per cent deployment), a low mobilisation (66.6 per cent deployment) and a high mobilisation (75 per cent deployment). The mid-range assumption has been chosen to demonstrate a probable rather than a worst-case deployment.

27. US Department of Defense, *Soviet military power 1984*, 3rd edn (US Government Printing Office, Washington, DC, 1984), p. 25.

28. NATO figures from *Jane's fighting ships 1985–86*, pp. 614, 672-3.

29. James John Tritten, *Soviet Navy data base*, pp. 30-3, 76, fns. 31-3.

30. Paul H. Nitze et al., *Securing the seas*, p. 109.

31. Lucas and Preston, 'USSR's Navy exercise', p. 547. Air Commodore G.S. Cooper, 'Soviet naval exercise', p. 4. Peter Almond, 'Soviet Fleet maneuvers "seize" eastern Atlantic', *Washington Times*, 24 July 1985, p. 1. Anon, 'Soviet Navy exercise to counter NATO', *JDW*, 27 July 1985, p. 155.

32. For a discussion of the lack of clarity over US submarine force levels to accomplish various missions, see CBO, *The US sea control mission,*

pp. 40–8. For an analysis of the number of submarines needed to man the GIUK Gap, see CBO, *Shaping the general purpose navy*, Appendix B, pp. 127–40.

33. James John Tritten, *Soviet Navy data base*, p. 30.

34. RUSI, *News brief 56: August 1985* (RUSI, London, 1985), pp. 4–6. The 18 open-ocean warships deployed in the July 1985 exercise were probably close to the Northern Fleet's maximum deployment. Of the remaining 21 open-ocean warships, five are old gun cruisers and destroyers only suitable for supporting amphibious landings. This leaves only 16 modern warships from which the Northern Fleet was unable, in the July 1985 exercise, to spare a proper escort for the nuclear-powered cruiser *Frunze*. A ship of such importance has, since 1983, been escorted by between five and seven warships (inter-fleet transfers excluded), but in July 1985 the *Frunze* sailed from Kola with only three escorts in company.

35. For example, see Robert Hutchinson and Antony Preston, 'Port mining threat launches new look at reinforcement plans', *JDW*, 14 January 1984, pp. 3–7. The new US Navy Maritime Strategy may well make US carrier deployments less cautious and more aggressive. See note 14 for references to the US Maritime Strategy.

36. Viktor Suvorov, 'Strategic command and control: the Soviet approach', *IDR*, no. 12 (1984), pp. 1813, 1817. Tomas Ries, 'Defending the far north', p. 877. The exact nature of the highest levels of command relationships is unclear. In particular it is not known whether the Navy keeps an essentially independent chain of command from the General Staff, or how such a command chain relates to the GTVD/TVD and oceanic TVD structure. For one possible command chain, see IISS, *The military balance 1985–86*, pp. 25–30, though this offers an incomplete picture.

37. Assumptions on expected hull-life are taken from CBO, *Shaping the general purpose navy*, pp. 1-2, fn. 2.

8

The Nordic Response to the Soviet Presence

Clive Archer

The Nordic countries have never had a uniform approach to the presence on their borders of a preponderance of power. In recent times the states of Denmark, Finland, Iceland, Norway and Sweden have adopted different security policies and, while politicians in each country have usually placed an emphasis on the need to maintain a consensus on the broad outlines of foreign and defence policies, there are dissident voices within each country that wish for alternative policies.

The five Nordic countries have adopted security policies since 1945 which have been both different from each other and yet have shown elements of overlap. During the Second World War, Norway and Denmark had been occupied by Nazi Germany; Iceland, Greenland and the Faroes had had a more benign occupation by British and American troops; Sweden had maintained a precarious armed neutrality; and Finland had fought three wars — two against the Soviet Union and one against Nazi Germany. In the immediate post-war period, the small states in northern Europe all hoped for a peace based on the wartime coalition of the Great Powers underpinning a system of collective security run by the United Nations. The onset of the Cold War in 1947-8 disappointed these hopes. The fall of the neutral, bridge-building democratic government of Czechoslovakia to a Communist Party *putsch* and the Soviet request to Finland for a Treaty of Friendship and Mutual Assistance in Easter 1948 frightened the governments of the Nordic states. The Danish, Norwegian and Swedish governments discussed the possibility of a Nordic Defence Union, but when it became clear that this was not a practical alternative, Norway and Denmark — joined by Iceland — signed the North Atlantic Treaty and Sweden remained outside any of the alliances.

Since 1949, the Nordic countries have maintained the broad patterns of the security policies they adopted in the 1948–9 period, though they have naturally undergone some refinement. Finland signed a Treaty of Friendship, Co-operation and Mutual Assistance with the Soviet Union in 1948 which committed it to preventing any attack on the Soviet Union through Finnish territory by Germany or its allies. The Treaty also required the Finns to call upon Soviet assistance in such a case or, indeed, if it were felt that such an attack were threatened. The Treaty established that Finland was to be regarded as a non-aligned country. Since 1948 successive Finnish presidents and governments have established a special relationship with the Soviet Union whereby Soviet confidence in Finnish neutrality — and in Finland's determination not to allow others to threaten the Soviet Union through Finland — is maintained by diplomacy. Sweden has continued its policy of staying out of alliances and also having a strong national defence. Its governments have also tried to encourage an international environment within which small, non-aligned, peaceful states such as Sweden find it easier to exist. They have contributed forces to UN peacekeeping operations and have been active in the pursuit of international arms control and disarmament. Norway, Denmark and Iceland have continued their membership of NATO and the first two have played their part in the integrated command structure of the Organization.

While the security policies of the five Nordic states have gone down different roads since 1948, the policy of each Nordic government is carried out with due regard to the interests of all the other Nordic countries. There is a Nordic dimension to the security policies of Denmark, Finland, Iceland, Norway and Sweden. It is not one of a united approach – except in certain UN matters – or even of convergence, but has traditionally been one of respect for each other's policies. Furthermore, all five countries have in common a balance between deterrence and reassurance in their security policies. The deterrence element is made up of the military strength of each country together with the assistance that may be made available from outside by virtue of alliance — in the case of Denmark, Iceland and Norway — or of a treaty of assistance, in the case of Finland. The reassurance side of the policies involves their diplomatic effort to show that their territory cannot be regarded as a source of insecurity for neighbours. It can be seen in the restraints exercised in Danish and Norwegian defence policy, in the non-alignment stance of Sweden and Finland and in the special relationship of Finland with the Soviet Union.

During the period when Denmark, Norway and Iceland were negotiating membership of NATO, opposition to joining a military alliance was voiced in all three countries. Though it was most powerful in Norway, even there it was clearly a minority viewpoint.[1] Prudential considerations of the trauma involved in such a change from non-alignment to a commitment to NATO and of the effect of such a move on the Soviet Union's treatment of the Nordic region led the Norwegian government (later followed by its Danish counterpart) to commit itself unilaterally to a policy forbidding foreign bases on its territory. In a note to the Soviet Union (which had warned Norway against membership of NATO), the Norwegian government stated that it

> ...would not enter into any agreement with other states which involve obligations for Norway to establish bases for the military forces of foreign powers on Norwegian territory, so long as Norway is not attacked nor subject to the threat of attack.[2]

This was both a reassurance and a reminder. It was intended to confirm to the Soviets that its close neighbour, Norway, would not — under normal circumstances — allow other countries to threaten Soviet security by establishing bases on its territory. At the same time, the Norwegians made it clear that if Norway's existence were in jeopardy, this policy of restraint could be reversed.

The same mixture of restraint and reservation was attached to Norwegian and Danish official attitudes towards the emplacement of nuclear weapons on their soil. Whilst subscribing to the nuclear strategy of NATO, Norwegian and Danish governments were unhappy about having atomic weapons on their soil, partly because of the hostility of their people to such weapons and partly because they felt that such weapons would upset European security if placed so close to Soviet territory.[3] The original refusal of Honest John and Nike nuclear-armed, short-ranged missiles in the 1957–60 period led to the refinement of a 'no nuclear weapons' policy by both governments. The Danish formula typifies the approach: they have declined to have nuclear weapons on their soil for their own use or that of allies 'under existing/present circumstances/conditions/ situation'. The Social Democrat Foreign Minister, Kjeld Olesen, used a variation of these words in October 1980 and added 'which means peacetime'.[4]

These are not the only self-imposed restraints that the Nordic NATO members have adopted. Norway has not allowed Allied troops, airplanes or ships to exercise near its northern border with the Soviet Union[5] and Denmark limited such exercises on its Baltic island of Bornholm. Iceland has no indigenous armed forces, the Iceland

Defense Force being provided by the United States.

It is important to place these restraints in perspective. They are all ones which are self-imposed by the governments of the countries and, as such, can be unilaterally reversed. Indeed, it is partly the prospect of such changes that gives them their strength: any *coup de main* by the Soviet Union in the Nordic region could be counterproductive if it brought with it a reversal of any of these restraints. Secondly, none of these policies detract from the full membership of Denmark, Iceland and Norway of NATO. All three are original signatories of the North Atlantic Treaty and Halvard Lange, the then Foreign Minister of Norway, could be said to be a Founding Father of the Alliance. Denmark and Norway provide conscript armed forces which form part of NATO's system of deterrence and defence, and Iceland hosts an important NATO base at Keflavik. All three countries play a substantial role in NATO's command, control and communications structure and provide important territory for reconnaissance and surveillance of Soviet military activity. A third point is that much of the restraint by the Nordic NATO members is a recognition that the Soviet Union has strategic resources near their territory (for example, on the Kola Peninsula or around Leningrad) which are not primarily a threat to the Nordic region but which are part of the East–West strategic balance and which should not therefore be directly threatened from Norwegian or Danish territory. It is recognised that to do so would only lead the Soviet Union to direct its might against such a threat and could also upset the strategic balance. Implicit in this policy of restraint is the hope that the Soviet Union will respond to the Danish and Norwegian self-denying ordinances by not allowing their forces near the Nordic region to appear to be a distinct threat to the countries of that area. Questions arise as to whether the build-up of Soviet forces in northern Europe constitutes a threat to the security of the Nordic states and, if so, as to what the response of those countries has been.

The first question is difficult to answer. There have been those — such as Nils Ørvik[6] — who have warned for some time that the growth of Soviet military power in the north represents a direct threat to Scandinavia. Two authors in this volume, van Tol and Ries, seem to point to wider Soviet considerations in their land, sea and air deployments in northern Europe.[7] Whatever the motivation of the Soviet build-up, it is bound to have an effect on the feeling of security of the Nordic states, especially when taken together with American deployments in Europe and in Northern Waters and the superpower tension of the late 1970s and early 1980s.

On the subject of the Nordic countries' responses, it is important to distinguish between the official attitudes and those of the wider public, and between the military and the political responses. In the account below, particular attention will be paid to Denmark, Iceland and Norway as these are the three Nordic states with direct interests in Northern Waters.

Denmark's main security interests lie in the Baltic, the approaches to which Denmark controls.[8] Their concern for Northern Waters is brought about by two outposts of the Danish kingdom — the Faroes and Greenland — being located in the North Atlantic.

The military response by Denmark and its NATO allies to an increased Soviet presence in Northern Waters has been to modernise and extend the installations in the Faroes and Greenland.

The military installations on the Faroes before the 1970s consisted of a Loran-C navigation station established at Ejde in 1959 and the airforce station at Mjørkadal, opened in 1962, which also hosted a NATO Early Warning Station. Furthermore, there was a radar station at Sornfelli which acted as a communications link between North America and Western Europe. A similar link went through the Faroese Command headquarters at Thorshavn.[9] In 1961 the Faroese assembly expressed its opposition to the storage of war materials on its country's land and sea territory. Since 1980 installations on the Faroe Islands have been updated but this process has caused a certain amount of adverse commentary in the islands.[10]

Since April 1941 there have been American bases (now called 'defence areas') on Greenland. In 1951 a bilateral agreement was signed between Denmark and the United States, which established three joint US–Danish defence areas under local American command at Thule, Søndre Stromfjord and Narssarssuaq. Narssarssuaq was eventually handed back to Denmark in 1958, but the other two areas were used as important US Air Force stations. Thule was originally utilised as a staging-post for the Strategic Air Command's (SAC's) bombers and refuelling tankers and later was developed to house Site I of the US Ballistic Missile Early Warning System (BMEWS). Thule also acts as a relay station for meteorological, navigation, communications and surveillance satellites. Sondrestrom Air Base was used as a dispersal airfield by the US North-east Air Command and then by SAC. From 1960 it also ran the Distant Early Warning (DEW) line stations in Greenland — DYE 1, DYE 2, DYE 3 and DYE 4.[11]

Since the 1970s, the two bases have been developed in line with the need to protect North America from Soviet incursions and to link North America with Western Europe. Thule Air Base has been taken over by the US Space Command and has a greater space surveillance task than previously. The BMEWS is being updated for this duty and there are facilities at the base for communications with a range of US satellites and with US military airplanes. Since 1981, increased Soviet bomber activity in Northern Waters, with the prospect of air-launched cruise missiles being carried by such aircraft, has led the US to update its DEW-line. As in the Faroe Islands, the development of defence establishments in Greenland has come in for some local criticism. Concern has been shown lest the new arrangements make Greenland part of what has been described as 'vital elements of an offensive first-strike capability' by the United States. This approach was one adopted by 'peace researchers' and left-wing politicians in Copenhagen, but some of the ideology and verbiage of the debate has been taken up by local politicians in Greenland. In particular, members of the two coalition governing parties questioned the handling of defence issues relating to Greenland by the chairman of the local assembly, Jonathan Motzfeldt. They considered that he had accepted too easily American assurances about the non-provocative nature of the BMEWS update and had kept facts from other members of the Greenlandic executive. This disagreement led to a snap election in May 1987 which returned the coalition, strengthening left-wing nationalist elements and leaving the whole issue unresolved. However, despite this intrusion of a defence-related issue into the Greenlandic political scene, the US bases have a low salience in local politics there when compared with social and economic issues and with fisheries and development policies.[12] There has been some indication that Soviet attempts to meddle in Greenland have been rebuffed.[13]

To summarise the situation regarding Denmark and the Northern Waters, it is fair to say that the increased Soviet presence in the area has not greatly impinged on Danish defence resources which are committed primarily to the Baltic area. Instead, the response in the North Atlantic has come through Denmark's NATO allies, primarily the United States. Denmark's contribution has been to make available facilities in the Faroes and in Greenland for US and NATO use. This has not been done indiscriminately, nor without control and consultation. As local politicians in the Faroes and Greenland have started to become interested in defence matters, so the Danes have widened the consultative process to include the relevant authorities in the two countries. In both the Faroes and in Greenland, the local assemblies — which do not have jurisdiction over defence matters — have voted to keep their countries nuclear weapon-free and the Danish government itself has been under certain pressure to participate in discussions for a Nordic nuclear-weapon-free zone. As the centre–right government in Denmark does not seem to have a parliamentary majority for its own defence policy, it finds itself in a difficult position. It should be remembered that any Danish response to the growth in Soviet forces in and around Northern Waters is going to be determined more by domestic political factors than by Soviet manoeuvres. With regard to defence matters in Danish politics, there is the feeling that 'it makes little difference what Denmark does'.[14] There is, however, the danger that Denmark's unwillingness or inability to improve its own military position in the face of an extended Soviet presence in Northern Waters and in the Baltic may cause its NATO allies to reconsider their commitment to reinforce Denmark in times of crisis. The United Kingdom government is already reconsidering the commitment to reinforce the Baltic Approaches with the UK Mobile Force, mainly because of high support costs, and both Britain and the Netherlands have indicated to Copenhagen that any radical change in Danish defence plans — such as those canvassed by the Danish socialists — would put at risk Allied help in a war.[15]

A country with a direct interest in strategic developments in the North Atlantic is Iceland. In 1951 Iceland agreed to the stationing of US forces there to defend the island and its surrounding air and sea area. This Iceland Defense Force was under national US control until 1986 when its status was changed to that of a NATO headquarters under SACLANT. The Keflavik base services two surveillance radars and AWACS airplanes are permanently stationed there. The base co-ordinates surveillance activities in Northern Waters, and submarine movements are tracked through the SOSUS chain off the Icelandic coast.[16] As elsewhere in the North Atlantic, facilities in Iceland have recently been updated. Eighteen F-15s have replaced the 12 F-4E Phantom interceptors and will be based in new semi-hardened aircraft hangars. The two existing radar stations are being modernised and there are plans for two extra stations in the Westfjord and north-east Iceland areas.[17]

This increased US–NATO activity has been matched by growing Icelandic involvement in NATO: the Icelandic Coast Guard is to be trained in minesweeping Icelandic territorial waters, and Iceland has plans to involve itself in NATO's Military Committee and to increase civilian manning of defence installations in the country.[18] This positive attitude towards security matters has been made possible by a change that removed the anti-NATO People's Alliance from government and led to a coalition of two pro-NATO parties which lasted until mid-1987.[19] In July 1987 these parties were joined in government by a third pro-NATO party. It was also encouraged by the work of an all-party committee, appointed in 1979 to report on security matters, and by the realisation by political leaders — made more aware of their position by the work of the committee — that 'the expansion of the Soviet Fleet, with its huge base on the Kola Peninsula, has made Iceland more vulnerable than ever'.[20] Public opinion has gone in tandem with the politicians' views: support for NATO within Iceland is now strong among the electorate.[21] Thus, there seems to be *prima facie* evidence that the increased Soviet presence in the North Atlantic has bolstered pro-NATO sentiment in Iceland and encouraged its government to take a more active role in NATO.

Some words of caution should be added. Until 1976 Iceland had a series of fishing disputes with a major NATO country — the United Kingdom — and, until 1974, the presence of the US base at Keflavik was an active political issue in Icelandic politics. Both questions produced negative views of NATO within sections of the Icelandic electorate. The 1974 agreement on the base between Iceland and the United States (and also the subsequent moves to multilateralise the base within NATO) and the 1976 settlement of the fisheries dispute with Britain, an outcome which a number of NATO countries facilitated, helped to remove for Icelanders two major political stigmas from NATO. Other issues have arisen in Iceland's bilateral relationship with NATO members — noticeably the United States — which have aroused public opinion. The questions of shipping materiel to the Keflavik base and US action against Icelandic whaling are two examples.[22] As yet there is little sign that these issues can be utilised to turn public opinion against NATO generally.

The European NATO member most affected by the growth of the Soviets' presence in and around Northern Waters is Norway. Norway is a 'front-line' state which has both a land and sea frontier with the Soviet Union. The key words for Norway's post-1949 security policy have been identified above as deterrence and reassurance. To be worthwhile the deterrence has to be credible and the reassurance reciprocated (or at least recognised). Events in the 1980s have given a certain edge to Norwegian security policy: it has had to adjust quickly to a variety of stimuli.

On the military side, Soviet activities have been perceived as a threat to Norwegian security in the far north. In February 1984, the Norwegian Chief of Defence, General Sven Hauge, pointed to the build-up of Soviet helicopters, particularly the combat type, in the army units in the Kola Peninsula as a factor reducing the advantages of Norway's troops.[23] Norway's then Minister of Defence, Anders Sjaastad, considered that, should the Soviet Union gain control over the Norwegian Sea in a war, large tracts of Norwegian territory would lie behind Soviet naval operational areas, making a common defence of Norway — especially the north — more difficult. He expressed disquiet about the USSR's naval expansion, particularly that into the open seas off Norway.[24]

Norway has adopted a two-pronged response to this threat: it has increased its own defence effort and has further secured Allied promises of outside aid. Part of the Norwegian response can be gauged by the increase in defence expenditure. Over the period from 1979 to 1983, Norway averaged an annual real growth in defence spending of 2.7 per cent, placing it fourth, on this issue, in the NATO-Europe league table.[25] Furthermore, the centre–right coalition in power from September 1981 to May 1986 planned a 20 per cent real growth in the defence budget for the five-year period of 1984-8 inclusive,[26] though the change to a Labour government and the depression in oil prices led to a downward adjustment of that figure. So far, much of the expenditure has gone into the purchase of F-16 aircraft and in starting the upgrading of the army's brigade structure. The priorities for the years 1984-8 are to be: national measures for pre-stocking, air defence (EYEHAWK) for air bases, further equipment improvements for army brigades, light air-defence for the army and navy, six new submarines, PENGUIN Mk 3 anti-ship missiles for the F-16, increased army ammunition stocks and up-to-date weapons for navy vessels.[27] However, Norway has not taken up the offer of extra F-16s to add to the 68 it has, and this has been explained in terms of lack of funds.[28] Most of the expenditure in the 1980s will aim at giving Norwegian forces the ability to survive the first shock of an attack and to go on to cover and support the reinforcement of the north both from within Norway and from outside.

Indeed, Norway hopes to stretch both its ground forces and air power in times of war or crisis by 'buying in' support from the NATO Allies. That is why it is so important for Norwegian forces to survive any initial onslaught: the Allies must feel it is worthwhile committing resources to Norway. Currently the US Marine Amphibious Brigade (MAB) with its important Air Wing and the British–Netherlands Amphibious Force are earmarked for deployment to AFNORTH, with North Norway being a likely option.[29] Another possible reinforcement is that by the Allied Command Europe (ACE) Mobile Force. However, the Canadian White Paper on Defence of June 1987 ended that country's commitment of dedicating the CAST Brigade to North Norway. This was clearly a setback for Norway's policy of reinforcement, though any ill-effects should be reckoned in political rather than in military terms. In order to encourage outside reinforcement, Norway has placed emphasis on providing the infrastructure for it and also on arranging pre-stocking of equipment with Allies.[30] Norway has concluded crucial Collocated Operating Base (COB) agreements that allow US air resources to operate from Norwegian airfields in wartime.[31]

Maritime forces of importance for Norway are SACLANT's Maritime Contingency Force Atlantic (MARCONFORLANT), consisting of Striking Fleet Atlantic as well as the amphibious forces, and Standing Naval Force Atlantic (STANAVFORLANT) which has a number of destroyers or frigates from various NATO members.[32] There has also been a continued emphasis on surveillance by the Alliance in Northern Waters. The UK Royal Air Force (RAF) and the Royal Norwegian Air Force are of particular relevance for the Maritime Patrol of the Barents Sea, Norwegian Sea and northern part of the North Sea. The Royal Norwegian Air Force is to receive four new Orion P-3Cs by 1989 to replace its five outdated P-3Bs on patrol duty in the Norwegian and Barents Seas. Norway has attached importance to expenditure for NATO's Early Warning and Control System (NAEW & C) and for the joint command, control and information system (NORCCIS), integrated into the Allied set-up.[33]

Thus, as far as the deterrence side of Norwegian security is concerned, the early 1980s saw a strengthening of both Norwegian and Allied military resolve in northern Europe and Northern Waters. However, there has been no let-up on the self-imposed restraints on military activities instituted by successive Norwegian governments. One response to the increased naval activities off Norway's northern coast has been the Ministry of Foreign Affairs' attempt to strengthen the political steering of Allied military exercises in the north.[34] This seems to be an attempt by the minority Labour government, which took over power in May 1986, to respond to some of the more robust statements made in the United States during the debate on the Forward Maritime Strategy. As such, it would represent an effort to maintain US interest in the defence of Norway whilst reassuring the Soviet Union that Norway would not be a party to any provocative action off her coast.

When turning to the political response to the Soviet presence in Northern Waters, it should be noted that it is difficult to determine to what extent political activities have been determined by domestic or external factors, or whether one particular event has affected political action more than another. What can be examined — though not always determined — is public opinion on a variety of security questions and particular political responses to certain issues.

It has been noted that public opinion in Iceland has become more pro-NATO in the first half of the 1980s. This seems to be a general trend in Denmark and Norway as well. In 1981, opinion polls showed that 58 per cent of a representative sample of voters in Denmark supported NATO membership (compared with 49 per cent in 1964 and 52 per cent in 1974), and 72 per cent of a Norwegian sample in Norway wanted continued NATO membership, the 1965 figure being 44 per cent, and that in 1973 67 per cent.[35]

However, neither public opinion nor the politicians have been uncritical in their support for the *status quo* in security policy. It is noticeable that the Danish public's support for its defence forces is matched by an unwillingness to vote more money for that effort.[36] There have been specific issues in each country over which public opinion has shown reserve. In Iceland, it has been the question of the US base at Keflavik;[37] and in Norway and Denmark, the NATO Dual Track decision on INF has attracted voter hostility.[38] It would seem that nuclear issues in particular have worried the person in the street in Norway and Denmark: in 1982 38 per cent of respondents to a poll in Norway said that nuclear weapons were the issue of greatest concern to themselves and their country, a high rating when compared with sentiment in other Western states.[39] However, Norwegians were not indiscriminate when apportioning blame: in 1984 54 per cent of those questioned considered that the Soviet Union's military build-up to be most responsible for current international tension (this was the highest positive score for answers to this question among the nine Western electorates sampled).[40]

The left-wing parties in all three Nordic NATO states have always been against their country's membership of the Atlantic Alliance, though this has not stopped them from supporting — or, in the case of Iceland, even participating in — governments that have been pro-NATO. The 1980s have seen centre–right governments in power in all three countries for much of the time and the NATO-supportive Social Democrats or Labour Party in opposition. In that position, these parties have been more susceptible both to outflanking on defence issues by parties further to the left and to the growth of anti-nuclear feeling within their own parties. There has also been a generational change with the old guard who remember 1949, the Korean War and even the Second World War, being replaced by those brought up on US anguish over Vietnam and who are sceptical about the current US brand of capitalism. The Labour Party in Norway and the Social Democrats in Denmark have both attacked the centre–right governments on specific security issues — INF, votes on nuclear-free zones at the UN and on support for SDI research — whilst giving their general support for NATO membership.

The most noticeable trend in public and political opinion in the Nordic states when it considered international security in the 1980s was the support given to the notion of a Nordic nuclear-free zone (NNFZ). The idea had previously been canvassed by Finnish politicians but, in October 1980, a Labour politician, Jens Evensen, brought the issue into Norwegian politics whence it spread to Denmark, Sweden and Greenland.[41] There is some indication that the support given by public opinion to the NNFZ plans reflected public concern with the perceived growing nuclearisation of Europe, not least the Kola and Baltic areas. However, most politicians in Norway and Denmark have considered that negotiations for a NNFZ should be seen in the context of East–West discussions, a view on the whole supported by public opinion. The public in Norway and government spokesmen in Sweden have also placed emphasis on the need for Soviet areas bordering on the Nordic region to be included in any commitments involved in a NNFZ.[42] It should be noted that declarations of 'nuclear-free zones' in Greenland and the Faroes do not have the force of law — not that these countries are in danger of being 'nuclearised' — and the May 1985 resolution by the Icelandic Parliament on the question of nuclear weapons on Icelandic soil was a restatement of existing policy.[43] The whole issue has been discussed by a series of meetings by Nordic ministers and parliamentarians but it may be the case that the occasion of these meetings became more important than their outcome.

In summary, it can be concluded that the Soviet military presence in the far north has evinced the response of a greater military effort in the region by Norway, Iceland, Denmark and their allies. It has also hardened public opinion to support NATO and to remain sceptical of the USSR. It has not had any marked effect on the detailed security debate in the Nordic countries which seems more determined by the interplay of wider international events and domestic political conditions. It is not practical to expect the Nordic states to increase their defence expenditure vastly because of the growth of the Soviet strategic presence in Northern Waters. This lack of willingness may *eventually* bring results undesired by the Norwegian and Danish decision-makers: it may discourage other NATO countries from committing their resources to the reinforcement of these states in time of crisis or war. This would undermine the basis of Nordic security since 1949 and plans for a Nordic nuclear-free zone could scarcely be expected to redeem the situation. It would emphasise the precarious nature of security in the northern region and might invite greater superpower presence. However, as yet, such a change seems some way off.

NOTES

1. See Haakon Lie, *Skjebneår 1945–1950* (Tiden Norsk Forlag, Oslo, 1985), pp. 378–427 *passim*, and Olav Riste, 'Was 1949 a turning point? Norway and the Western powers 1947–1950' in Olav Riste (ed.), *Western security: the formative years* (Universitetsforlaget, Oslo, 1985), pp. 132–6.
2. Utenriksdepartementet (Ministry of Foreign Affairs), *Om sikkerhet og nedrustning: Rustningskontroll-og nedrustningsarbeidets plass i sikkerhetspolitikken. St. meld. nr 101 (1981–82)* (On security and disarmament: the place of the work for arms control and disarmament in security policy) (Ministry of Foreign Affairs, Oslo, 1982), p. 28.
3. Det sikkerheds-og nedrustningspolitiske udvalg (The Security and Disarmament Policy Commission), *Dansk sikkerhedspolitik og forslagene om Norden som kernevåbenfri zone* (Danish security policy and proposals of Norden as a nuclear- weapon-free zone) (SNU, Copenhagen, 1982), pp. 72–3.
4. Ibid., p. 75.
5. Utenriksdepartementet, *Om sikkerhet*, p. 28. For an elucidation of these restraints, see letter from Colonel G. Gjeseth '"Eastern border" for allied planes', *Stavanger Aftenblad*, 28 February 1987.
6. Nils Ørvik, *Europe's Northern Cap and the Soviet Union* (Harvard University, Center for International Affairs, Cambridge, Mass., 1963; Occasional Paper no. 6).
7. See Chapters 6 and 7.
8. Forsvarskommandoen (The Defence Command), *Danmarks strategiske betydning* (Denmark's strategic significance) (Forlaget Europa, Copenhagen, 1984).
9. Henning Lund-Sørensen, 'Rapport om debatten fra 1970 til 1985 om de militaere installationer på Faerøerne' (Report on the debate from 1970 to 1985 on the military installations on the Faroes) in *Flådestrategier og nordisk sikkerhedspolitik. Bind 2* (Maritime strategy and Nordic security policy. Volume 2) (SNU, Copenhagen, 1986), pp. 9–10.
10. Ibid., pp. 14–24.
11. Clive Archer, 'Greenland and the Atlantic Alliance', *CENTREPIECE 7*, (Centre for Defence Studies, Aberdeen, Summer 1985), pp. 13–16; 'Valg i Grønland den 26 maj', *Nordisk Kontakt*, no. 5 (1987), pp. 27–8.
12. Archer, 'Greenland', pp. 22–56.
13. *Politiken Weekly*, 10–16 January 1986, p. 16.
14. Erling Bjøl, *Nordic security*, (IISS, London, 1983, Adelphi Papers no. 181), p. 34.

15. *Hansard, House of Commons*, Written Answer, 20 January 1987, col. 547, and *Strategic survey 1986–1987*, (IISS, London, 1987), p. 102.

16. *Flådestrategier og nordisk sikkerhedspolitik. Bind 1* (Maritime strategy and Nordic security policy. Volume 1) (SNU, Copenhagen, 1986), pp. 60–2.

17. Bjørn Bjarnason *News from Iceland*, June 1985, p. 21 and *Nordisk Kontakt*, no. 9 (1985), p. 598. See also *Utanrikismál: skyrsla Geirs Hallgrimssonar utanrikisradherra til Alpingis 1985* (Foreign Affairs Report: presented in Parliament by Foreign Minister Geir Hallgrimsson, 1985), (Gutenberg, Reykjavik, 1985), pp. 26–34.

18. *News from Iceland*, May 1985, p. 15 and March 1985, p. 9; *Nordisk Kontakt*, no. 6, 1987, pp. 49–50.

19. Bjørn Bjarnason 'Iceland's security policy' in J.J. Holst, K. Hunt and A. Sjaastad (eds), *Deterrence and defense in the north*, (Norwegian University Press, Oslo, 1985), pp. 142–4.

20. Ibid., pp. 148–9.

21. *News from Iceland*, August 1984, p. 2 and Olafur Hardason, *Icelandic attitudes towards security and foreign affairs*, (Icelandic Commission on Security and International Affairs, Reykjavik, 1985), pp. 9–15.

22. *News from Iceland*, September 1985, pp. 1 and 15, October 1985, p. 32, August 1986, pp. 1, 21, 30, 31, and March 1987, p. 32.

23. *Norinform*, 28 February 1984, p. 2.

24. *Norinform*, 26 March 1985, p. 5.

25. *House of Commons Third Report of the Defence Committee, Session 1984–5: Defence Commitments and Resources and the Defence Estimates 1985–1988 Vol. II*, (HMSO, London), p. 7.

26. *Norinform*, 28 February 1984, p. 2.

27. *St.meld.nr.74* (English translation), pp. 58–9, and in *NATO's sixteen nations*, June–July 1983, p. 131.

28. *Norinform*, 28 February 1984, p. 3.

29. See Sir Richard Lawson, 'Vi må holde stillingen', *Norges Forsvar*, nr. 1 (1985), pp. 10–11; Tønne Huitfeldt, *Nordflanken sett fra Brussel*, (DNAK, Oslo, Atlanterhavskomitéen Serier, nr. 80, 1983), and chapters by Michael Leonard and Tønne Huitfeldt in Holst, *et al.*, (eds) *Deterrence*.

30. *St.meld.nr.74* (English translation), pp. 22 and 24.

31. *Budsjett-innst.S.nr.7 (1983–84)*, pp. 20–5.

32. IISS, *Militaerbalansen 1984–85, Norsk utgave ved Den Norske Atlanterhavskomité*, (DNAK, IISS, Oslo, 1984), pp. 110–11.

33. Ibid., p. 56, and *Jane's Defence Weekly*, 20 September 1986, p. 599.

34. *Norinform*, 17 June 1986, p. 1.

35. Figures cited in M. Heisler 'Denmark's quest for security' in G. Flynn (ed.), *NATO's Northern Allies*, (Croom Helm, London, 1985), p. 75 and in J.J. Holst, 'Norwegian security policy' in Holst *et al.*, (eds.), *Deterrence*. p. 237.

36. Heisler, 'Denmark's quest', pp. 83–5.

37. See Hardason, *Icelandic attitudes*.

38. Heisler, 'Denmark's quest', p. 93 and Holst, 'Norwegian security policy', pp. 226 and 239. It should be noted that there has not been a consistent majority against deployment in either state.

39. Figure cited in Holst, 'Norwegian security policy' p. 233.

40. ibid.

41. C. Archer, 'Deterrence and reassurance in Northern Europe', *CENTREPIECE 6*, (Centre for Defence Studies, Aberdeen, Winter 1984), and also Archer, 'Greenland' for information about Greenlandic interest in the question.

42. Archer, 'Deterrence', and Holst, 'Norwegian security policy', p. 232.

43. Bjarnason, 'Iceland's security policy', p. 145.

9

Responding To The Soviet Presence In Northern Waters: An American Naval View

Douglas Norton

THE SETTING AND US INTERESTS

The extent of Northern Waters is defined for purposes of this book as from latitude 60° to 80° North and from longitude 90° West to longitude 40° East. A naval view is affected by considerations beyond these boundaries, including not only other sea areas but also land. A naval view therefore considers Northern Waters in relation to the Atlantic Ocean, North Sea, Baltic, Mediterranean and NATO's Northern, Central and Southern Regions. I will describe issues from that perspective.

If that is one point of departure — that Northern Waters should not be considered in isolation — here is another: a Soviet maritime presence in Northern Waters is not *ipso facto* remarkable, or a cause for either concern or a US reaction. These are, by and large, international waters and much of the area is close to the Soviet Union.

What is remarkable is the rapid development of the Soviet Union into a maritime power which, alone, now rivals all of NATO in this field. What is cause for concern, and reaction, is NATO's potentially crippling vulnerability to the world-class Soviet Navy if it is dispatched from its bases, the largest of which is within Northern Waters, to establish control of the Norwegian Sea and to sever the Alliance's sea lines of communication (SLOC) by intimidation in a crisis or by force of arms in war. It is this prospect to which the United States and its NATO Allies have reacted.

The surface, depths and skies of the Norwegian Sea, the shallow seas to its east and south and the Arctic waters to its north are a factor in military planning, since they constitute an important manoeuvre area enfolding NATO's Northern Region: Denmark, Iceland, Norway and their island territories. Within that Region Iceland, Greenland, Norway and the sea bearing its name can be considered as the Northern Flank. The Norwegian, North and Baltic Seas have long been 'home waters' for the maritime forces of Belgium, Britain, Denmark, Germany, the Netherlands and Norway.[1] As the maritime forces of the Soviet Union and Warsaw Pact grew, these seas also became home waters of their maritime forces. Eastern and Western maritime forces operate in and over them daily, going about the routine business of sea and air power in peacetime: signalling and supporting national interests, training, conducting ballistic missile submarine (SSBN) patrols and monitoring each other's activities.

Geopolitical factors such as Soviet power projected from the Kola Peninsula and the Leningrad Military District, Finnish and Swedish neutrality and the particular importance of reinforcement and resupply in defending the Northern Region would cast Northern Waters in a role of great significance in a crisis or during a conflict between NATO and the Warsaw Pact. Besides their role in defending NATO's Northern Region, the Norwegian Sea and the shallow seas surrounding the United Kingdom mingle in more than a literal sense with the Atlantic when one considers NATO's fundamental requirements for a successful defence. Among these rock-bottom requirements is a vast, transatlantic flow of materiel — 95 per cent of it shipborne — to continue the battle and sustain the people of the NATO alliance. Perhaps above all else, the seas and airspace extending west from the Baltic throat and south-west from North Cape would be a critical arena in a battle to maintain NATO's flow of supplies along SLOC.

The seas which wash the shores of NATO's Northern Region are not home waters of the United States. For the United States, Northern Waters in particular have none of the intrinsic interests, such as resource exploitation or access to the Atlantic or to island possessions, which concern the littoral nations. Despite the geographic reality which means, as Tomas Ries points out, that should strategic nuclear deterrence fail, the missiles and strategic bombers of the nuclear powers would arc above Northern Waters, as they would above several other areas half a world away from there, this possibility is not central to US military thinking about the area. There really are but two sources of significant, continuing US military interest in the sea areas north and east of the Atlantic Ocean proper: NATO obligations and monitoring Soviet forces.

The US government concerns itself on a regular basis with the plans, forces, and training required to support NATO in peace, crisis or war. NATO's Northern Region is a particularly demanding case since it would be especially dependent upon reinforcement to sustain deterrence and its capture and subsequent use as a Pact base of operations would gravely imperil the entire defence of the Alliance.

There are major concentrations of Soviet military power in the Kola and along the Baltic littoral. The rationale for monitoring them is self-evident. Just as do the other nations who share many of these same home waters, the Soviet Union conducts operational training in and over Northern Waters, the Baltic and the North Sea. The United States observes Soviet operations, as do the naval and air forces of NATO and neutral alike, and indeed, as the Soviet Union observes the maritime operations of others. It is a routine business with well-understood rules — many of which are codified in procedures agreed between the US Navy and the Soviet Navy.[2]

THE SOVIET BUILD-UP AND NATO'S REACTION

Other chapters describe in some detail the face of Soviet power on NATO's Northern Flank. That detail need not be repeated here, but we need to consider the political–military effects upon NATO of the Soviet military build-up in the Kola.

The magnitude, the context and the consequences of the growth of Soviet power to the north and east of what has been called 'Europe's quiet corner' have been described by Norwegian officials. A few days prior to taking up his duties in May 1986, the Labour government's Minister of Defence, Johan J. Holst, addressed a conference in London:[3]

> When Norway, Denmark and Iceland entered NATO as founding members, the Soviet Union did not constitute a serious threat to the trans-atlantic sea lines of communication (SLOC). However, through the nineteen seventies the Soviet naval buildup which appeared to signal an ambition to contest Western naval supremacy caused growing concern in Norway as it tended to cast doubts on the assumption that allied assistance could be brought to bear on a crisis in order to prevent the outbreak of war or, in the event of an actual attack, be interposed between the attacker and his objective in Norway. Allied assistance was dependent on transportation by sea and Soviet submarines and naval aircraft with standoff weapons emerged as a potent threat against its real effectuation in an emergency.

Another prominent Norwegian, Conservative Prime Minister Kaare Willoch, put the development of Soviet maritime power in the context of Western maritime trends. Speaking in 1986 in Oslo, Mr Willoch, then in office, said:[4]

> The expansion of the Soviet Navy should and must be viewed in the light of the Western reductions in naval forces during the same period. NATO countries now acknowledge, to an increasing degree, the necessity of counteracting the shift in the balance of forces which has taken place.

Perceptions of a relative decline in the ability of Norway's allies to assure reinforcement and thus effective defence of Norwegian territory, and the implications for NATO of Pact control of Norway in a conflict, caused Norwegians and their allies to search for ways to deal with the new reality. During the late 1970s Norway wrestled with the problem of how to react within the intricate balance of political–military factors in the Nordic region. Having a large territory and small standing military forces, Norway relies for deterrence and defence on a combination of rapid national mobilisation and prompt arrival of Allied reinforcements. After a lengthy national debate Norwegians chose several carefully calibrated measures in concert with Canada, the Netherlands, the United Kingdom and the United States.

These measures, described as a means of '...demonstrating resolve with the aim of deterring attack while limiting escalation potential', included plans to pre-stock equipment and strengthen reinforcements of defending aircraft: some heavy equipment to be pre-positioned in central Norway for the US Fourth Marine Amphibious Brigade (4 MAB), limited pre-stocking for the United Kingdom–Netherlands Amphibious Force (UK/NL AF) and for the Canadian Air–Sea Transportable (CAST) Brigade, plus a new rapid reinforcement plan for NATO fighter aircraft. Norway also decided to pre-stock equipment in the High North for another Norwegian Regimental Combat Team.[5]

More recently, Norway introduced first-line F-16 interceptors into its air force. Increased emphasis on naval readiness was highlighted by a no-notice naval exercise in 1985 when the entire navy was directed to battle stations at sea, a challenge to which it responded with a success rate to be envied by larger navies. In 1985 Canada assigned the reinforcement (CAST Brigade) task to an active formation rather than a reserve unit and agreed with Norway that in 1986 the full CAST Brigade would train there.

West Germany also observed the growth of Soviet power in the

Kola and responded with increased emphasis on the defence of the Northern Region. A Defence White Paper in 1979 specified that the Federal Republic has a vital interest in the security of NATO's key positions on the Northern Flank. In 1980 the German Federal Security Council lifted self-imposed navy operational restrictions and since then German naval operations above latitude 61° north and beyond the Dover–Calais line have increased. Today, as an official study puts it, 'The German Navy is an important instrument of forward defence at sea against attacks on key positions on NATO's Northern Flank...'.[6] A senior West German naval commander described forward defence at sea as a key element in deterrence of war:

> The NATO strategy of flexible response and a forward defence capability are the basis of deterrence. The forward defence strategy at sea does not confine naval and air forces to narrowly defined regional borders - it calls for their employment anywhere the situation dictates, even in peripheral seas where an adversary may attack.[7]

Another result of Soviet maritime power based in the Kola area is that '...Britain now tends to see the principal threat vector as much from the north as from the east'.[8] The 1985 White Paper on Defence observed that

> It must be assumed that only limited US Navy forces would be available in the Eastern Atlantic at the outbreak of hostilities. European navies, and in particular the Royal Navy must therefore be ready to play a leading role in initial operations.[9]

In 1977 NATO Defence Ministers determined that there was a need to assess the maritime situation, in view of the continuing importance of the maritime contribution to NATO. Ministers directed the NATO Military Authorities to develop a concept for Alliance maritime operations. Major NATO Commanders (the Supreme Allied Commander Europe — SACEUR, the Commander-in-Chief Channel — CINCHAN and the Supreme Allied Commander Atlantic — SACLANT) developed the Concept of Maritime Operations (CONMAROPS), a broad statement of how maritime forces may be used in support of NATO's basic defensive concept, Flexible Response.

CONMAROPS was approved by the nations of the Alliance in 1980 and revised in 1985. It gave visibility to the collective judgement of

NATO's senior commanders that Alliance planning to accomplish its basic strategic objective — the protection of members' territory — must take greater account of a Soviet maritime power now sufficient to pose a serious threat to NATO's SLOC. The broad outline of NATO maritime planning can be summarised as follows: to deter war with credible forces which are clearly ready to defend the Alliance; should deterrence fail, contain Soviet-Warsaw Pact maritime forces, tie them down in defensive tasks and neutralise those forces already deployed and prevent their reinforcement.[10]

The United States joined its European allies in responding to the situation created by the development of the Soviet Navy, and in particular the Northern Fleet, into a formidable blue-water force. Despite the demands of world-wide commitments, including increasing concerns about Soviet maritime power in the Pacific and security in Central America, the US Navy sends major warships to NATO's regular exercises in the Northern Region, such as OCEAN SAFARI and NORTHERN WEDDING. In addition, naval commanders are giving greater attention to practising tactics suited to the region's environmental conditions, its geography–hydrography and the new Soviet capabilities on NATO's Northern Flank. Fleet operations planners are searching for the resources to respond to Norwegian suggestions[11] that there be an increase in predictable, routine deployments to the Norwegian Sea.

Another important way that the US has joined in sharing Allied concerns and in responding to them has been in planning for the use of maritime force in deterrence and defence. In the early 1980s, when called upon to refine and publish a US Navy Maritime Strategy, planners in the Office of the Chief of Naval Operations took account of the principles and judgments of CONMAROPS.

THE US NAVY MARITIME STRATEGY

- **Genesis and context of Maritime Strategy**

The roots of today's US Navy Maritime Strategy are decades deep. The strategy is not a transient, political document or an idea centred in the Pentagon E-ring, as the use of the term 'The Lehman Doctrine' by some Northern Waters conferees (see p. xvii above) might imply. Norman Friedman expressed this thought clearly:

> It must be stressed that the maritime strategy reflects concepts

which have long been understood within the US Navy; it is neither a radical innovation nor a series of prescriptions imposed from above. It has, however, three new features. First, it is *explicit*. Second it is a choice from among many ideas, not merely a bland combination of all existing concepts of operation. Third, and perhaps most important, it is intended to be a long-term choice. The question which must now be posed is not whether the fundamental concept is tenable, but what can be done on a tactical or material basis to implement the strategy.[12]

In considering what that strategy means to Northern Waters, it is necessary to keep in view the strategic context in which it is set. That context has four other elements: US national military strategy, NATO's strategy of flexible response and forward defence, the NATO CONMAROPS and the global elements of the Maritime Strategy itself.[13]

The national military strategy of the United States is deterrence and its purpose is to preserve US and Allied independence, integrity and freedom, and our vital interests. The United States seeks to achieve these objectives first without war, but if deterrence fails, by fighting to defend them and restore peace.

To carry out the national deterrent strategy US military forces must be able to perform key missions including nuclear deterrence; defence of vital interests in NATO, North-east Asia and South-west Asia; and protection of SLOC and power projection. The United States believes that deterrence is strengthened by forward deployment of our forces and with the concurrence of Allied governments has deployed ground and air forces in Europe, Japan and Korea. To this same end, US Navy carrier battle groups and amphibious forces deploy regularly to the western Pacific, Mediterranean and the Indian Ocean.[14]

NATO's military strategy is deterrence and if deterrence fails, flexible response and forward defence, which includes the option of escalation to nuclear strike. It was adopted by the Alliance in 1967 as a means of deterring overt attack, whether limited or general in scope, and of enabling the Alliance to continue to resist intimidation and coercion by the Soviet Union and its allies. General Bernard Rogers, then SACEUR, has described the range of military responses to attack provided by this strategy:[15]

- Direct defence, to defeat an attack or to place the burden of escalation on the aggressor. This is NATO's preferred response.
- Deliberate escalation on NATO's part, to include possibly the

first-use of theatre nuclear weapons.
- General nuclear response, the ultimate guarantor of Alliance deterrence.

A reference book published by NATO authorities in Brussels describes the intent and requirements of the Flexible Response strategy:[16]

> The NATO policy and strategy designed to deter aggression and to prevent war is based upon the capabilities of its conventional and nuclear forces and the evident intent and expertise on the part of Alliance political decision-makers and military leaders to use them, as necessary, in a timely and flexible fashion, should deterrence fail.
>
> This leads to the maintenance of a strong, diverse and flexible nuclear posture. Furthermore, to enhance the credibility of the deterrent, the member countries of NATO must all share the risks of escalation. NATO's basic doctrine and related posture is designed to convince potential enemy leaders at any time that the risks involved in initiating a war against NATO would be out of all proportion to any conceivable gains.

In another section of the same document, the roles of NATO's naval forces are said to be '...neutralising Soviet strategic nuclear submarines, safeguarding trans-atlantic sea lines, and in general preventing the Warsaw Pact from gaining maritime supremacy in the North Atlantic...'[17]

CONMAROPS describes maritime support of NATO's strategic concept in terms of the major campaigns to be expected. These are the Norwegian Sea, Shallow Seas (the Baltic, North Sea and approaches to the British Isles), Atlantic and Mediterranean.[18] These areas fall within an arc of interrelated sea areas in which operations vital both to the SLOC and to support of the land battle must be conducted. SACLANT has reported that he is about 50 per cent short of the maritime forces he would require to carry out these campaigns simultaneously.[19] NATO will have to set priorities in defending its territory and resources.

Maritime support of the land battle requires above all protecting the flow of cargo across the Atlantic and shallow seas into the ports of Europe. CONMAROPS posits that the key to winning the Battle of the Atlantic is winning the Battle of the Norwegian Sea and describes several critical relationships: the loss of northern Norway would be a determining factor in the Battle of the Atlantic as would the loss of Iceland; the loss of Greenland would be severe; losing control of the

Baltic Straits would increase hazards in both the ports and maritime approaches to Europe and in the Norwegian Sea.[20] CONMAROPS establishes a difficult decision matrix for NATO authorities, given the shortage of forces, because it also describes the requirement for strong maritime support, particularly air support, of the battle ashore in NATO's Southern Region.

- **The Maritime Strategy: deterrence is the goal**

The main features of the US Navy Maritime Strategy were set forth in January 1986 by the then Chief of Naval Operations (CNO), Admiral James Watkins, the Commandant of the Marine Corps, General P.X. Kelley, and Secretary of the Navy, John Lehman.[21]

The Maritime Strategy is the maritime component of US national strategy. It is deterrent in nature, reflecting the recognition of Navy leadership that peacetime operations and response in crisis are crucial contributions to deterrence and stability and, indeed, are equal in importance to the ability to fight if deterrence fails. Critics of the Maritime Strategy frequently ignore this point. Admiral Watkins wrote:

> One key goal of our peacetime strategy is to further international stability through support of regional balances of power. The more stable the international environment the lower the probability that the Soviets will risk war with the West. Thus our peacetime strategy must support US alliances and friendships...through a variety of peacetime operations including naval ship visits to foreign ports and training exercises with foreign navies.[22]

> The heart of our evolving Maritime Strategy is crisis response. If war with the Soviets ever comes, it will probably result from a crisis that escalates out of control. Our ability to contain and control crisis is an important factor in our ability to prevent global conflict.[23]

US Navy strategists believe that the Soviets want to avoid a superpower military confrontation but that localised conflicts and crises in the Third World will continue. The Maritime Strategy notes that

> a fundamental component of the nation's success in deterring war with the Soviet Union depends upon our ability to stabilise and control escalation in Third World crises...(the) Navy devotes much

of its effort to maintaining...stability. Potential crises and the aftermath of crises have increasingly defined the location and character of our forward deployments..., a continual presence in the Indian Ocean, Persian Gulf and Caribbean as well as our more traditional forward deployments in the Mediterranean and the Western Pacific.[24]

Just as do US national military strategy, NATO's Flexible Response strategy and CONMAROPS, the US Navy Maritime Strategy recognises that deterrence could fail and it considers, as the other strategies do, how the Navy would be used in a global war with the Soviets. This is, of course, essential to a deterrent strategy: a nation — or an alliance — which practises a deterrent strategy has clearly and credibly to communicate to a potential attacker that consequences which he finds unacceptable would follow its aggression. We may safely believe that if the Soviet Union should start a European war, Soviet leaders would prefer that it remain non-nuclear, confined to Europe and indeed, confined at any one time to the nation(s) actually attacked. The provisions of the North Atlantic Treaty, coupled with NATO's Flexible Response strategy, deter by stating that such a war would not be confined to Europe, would involve 16 democratic nations, and could well become nuclear.

The US Navy Maritime Strategy is firmly set in that context and supports NATO's military strategy. It does so by addressing itself to ways in which the preferred Soviet strategy for a European war may be countered if deterrence fails by global, coalition warfare, which could include the destruction of Soviet forces world-wide, attacks on Soviet flanks, not only in Europe but in the Far East, and creating a trend in the correlation of nuclear forces unfavourable to the Soviets. The global outlines of the Maritime Strategy must be kept in view during any consideration of the strategy on a lesser scale, because global maritime requirements will always be a factor bounding possible courses of action in any one area, such as NATO's Northern Region.

The strategy considers that at some point in a deepening superpower confrontation considerations of deterrence and defence would call for co-ordinated, additional world-wide US naval movements. The goal of this first phase of the maritime strategy is deterrence: sustaining US and Allied interests without war.

The majority of these movements would be dispatching additional forces to European and Asian waters because that is where our maritime forces must be to extend deterrence to our allies, protect

SLOCs and support the land battle if deterrence fails. (However, the reach of Soviet maritime power into waters such as the Caribbean requires that we do not denude those areas.) Among these movements would probably be, subject to mutual agreement, the initial elements of US Marines committed to NATO for the reinforcement of Norway's defences.

From the perspective of deterring war in Europe, the most important single act would be NATO's initiation of SACEUR's Rapid Reinforcement Plan. The global forward movement of US maritime forces would complement that strengthening of deterrence ashore by making clear that we will not leave allies who find themselves along the littoral of Soviet home waters to bear the burden of deterrence alone. It would also, in concert with the position of our deployed Air Force squadrons, signal that if the Soviets and their allies attack Europe, the Soviets will have to face US forces in the Far East and North Pacific as well.

These steps must be initiated promptly because, for example, an amphibious group at sea near Norfolk, Virginia, would take about nine days to reach southern Norway and a carrier group about ten days to reach the Mediterranean. Ships at sea near San Diego, California, would require about nine days to reach the western Pacific and three weeks to reach the Indian Ocean. These times assume that the troops, ordnance and supplies were already loaded aboard. Several additional days would be required if they were not.

- **The Maritime Strategy: if deterrence fails**

If war begins, the strategy, like CONMAROPS, calls for US and Allied forces to seize the initiative, destroy forward-deployed Soviet forces and move towards Soviet home waters. In this situation, which may be thought of as the second phase of the Maritime Strategy, the goal of maritime operations is to contribute to the rapid termination of war on terms acceptable to the United States and Allies. In the CNO's words:[25]

> To accomplish this maritime forces must counter a (Soviet) first salvo, wear down enemy forces, protect sea lines of communication, continue reinforcement and resupply and improve positioning. We must defeat Soviet maritime strength in all its dimensions, including base support.

One of the most important requirements during this period is protecting transatlantic shipping, which could otherwise take disastrous losses under attacks by the large, modern Soviet submarine force and by the missile-armed bombers of Soviet naval aviation. Carrier battle forces (groups of carriers, escort ships and supporting submarines) are central to achieving the goals of the maritime strategy and of CONMAROPS. These forces, which of course are subject to attack from the same sources as are reinforcement and resupply shipping, would need to attain and sustain themselves in positions where their aircraft and cruise missiles can support NATO's defence on the Northern and Southern Flanks, and they could be needed near the UK area and Central Region as well. Pacific Fleet carriers would be an important component of US military pressure on the Soviets in North-east Asia. If not required in North American waters, US Navy mine warfare forces would join their more numerous Allied counterparts in both mine-clearing and mine-laying operations.

After enough of Soviet maritime power had been destroyed to reduce sharply its offensive capability against the flanks of the land battle and against the SLOC — in classical terms, after the destruction of the Soviet 'fleet in being' — the tasks set by the maritime strategy become those of carrying the fight to the enemy. How this capability would be employed, which options would be selected, would depend on many factors. SACEUR's success in holding the Warsaw Pact first echelon and blunting the thrust of follow-on echelons would be a critical consideration.

As the forces of Allied Command Europe (ACE) sustained their defence and began to inflict reverses on the invading armies, maritime power could be employed to: keep up global pressure on the Soviets; complete the destruction of the Soviet Navy; support the land battle with air and amphibious attacks; pose a threat to the Soviet littoral; and influence the Soviet calculation of the correlation of nuclear forces in a way which, considered in conjunction with failure to achieve a rapid victory on the ground, could influence them to terminate hostilities.

The operative effects of the Maritime Strategy, in addition to enhancing deterrence, are threefold: the strategy offers a global perspective to the navy's operational commanders and a framework for advice to the National Command Authorities (the President and the Secretary of Defense). It thus gives a sense of the recommendations which CNO and the Navy Secretary, who are not commanders, might make to those who are. Such recommendations could then become tasks for navy operational commanders. The Maritime Strategy is also a key element in decisions about the kinds of quantities of equipment and people the navy has. Equally, it is a means for the navy's leadership to shape service thinking about probable missions and what sort of training is necessary to accomplish them.

Some Northern Waters conferees (see p. xvii above) spoke of the Maritime Strategy as if it were equivalent to the detailed military plans of the early 1900s, which many contemporary observers believe were so action-forcing and inflexible that in the crisis of 1914 they propelled Europe over the brink of war. I believe that a considered review of the Maritime Strategy, for which the summary provided here is clearly no substitute, will satisfy most observers that it is not a maritime Schlieffen Plan. It has no fixed timetables, it does not schedule the movement of aircraft carriers or other forces, it does not prescribe specific targets. It is a broad outline of ways in which US maritime power is used today, in peacetime, to sustain US and Allied interests and deter war, and of how the navy's top leadership views the role of maritime power in ending war rapidly on favourable terms if deterrence fails.

NORTHERN WATERS AS A THEATRE OF MARITIME OPERATIONS

- **Peacetime activities**

Northern Waters are a variable and difficult maritime environment, for example in terms of the undersea acoustic conditions critical to submarines and submarine hunters and of flying weather. Nations need to train seamen and aviators to deal with these conditions, and do so by means of routine operations in Northern Waters.

Periodically, NATO's maritime forces assemble for larger exercises such as OCEAN SAFARI and NORTHERN WEDDING and operate together in the North Atlantic, Norwegian Sea, North Sea and Baltic. Such regular exercises within and near Northern Waters enhance deterrence and add an important additional dimension to the benefits of national training: practice for NATO commanders and their staffs in NATO defensive plans, developing teamwork between the units of different countries and exercising NATO command and control procedures. (Land–maritime–air interfaces are particularly complex in NATO's Northern Region.) Other demonstrations of NATO solidarity and deterrent power are found in the form of periodic Northern Region deployments of the Standing Naval Force Atlantic (STANAVFORLANT), a group of about five destroyers or frigates, and of mine warfare vessels of the Standing Naval Force Channel (STANAVFORCHAN).

Though not maritime forces, two multinational NATO units with deterrent functions train regularly in the Northern Region: NATO Airborne Early Warning Force (NAEW) multinational air crews operate NATO's own E-3 Airborne Warning and Control System (AWACS) aircraft on regular deployments from their base in West Germany to areas throughout NATO, including the Northern Flank. Also found regularly in this area are the land and air components of the Allied Command Europe Mobile Force (AMF). Among the deployment options for this on-call multinational deterrent force are Northern Region contingencies.

In addition to training, surveillance of Soviet military activity is an important peacetime task in its own right, throughout NATO no less than in and over Northern Waters. Since the nations of the Alliance rely heavily on time to mobilise reserves and to bring reinforcing troops, ships and aircraft from North America for deterrence in a crisis and defence if attacked, early awareness of major changes in Soviet military activities is critical. Thus, for example, if the Northern and Baltic Fleets go to sea in unusual numbers, NATO needs to know it as soon as possible.

- **In crisis: sustaining deterrence**

Should a European crisis develop, NATO's Northern Region, for which Northern Waters are a significant consideration, could be a focal point of deterrence. Let me stress at this point that I do not believe that Soviet military action solely against the territory of any one NATO nation is likely at present. 'Finnmark Grab' or 'Berlin' scenarios and other types of isolated incursions seem improbable so long as NATO retains the political cohesion it has demonstrated, for example, by the deployment of ground-launched cruise missiles and Pershing II missiles. The discussion which follows sketches some important aspects of sustaining deterrence in a major crisis between NATO and the Warsaw Pact.

In such a crisis the goal of the Secretary-general and his national counterparts, who collectively constitute the leadership of NATO, would be to deter a Pact attack while maintaining Western vital interests. Several aspects of maintaining deterrence in a crisis would involve Northern Waters and the seas and land adjoining them.

Surveillance of the Northern and Baltic Fleets is one such aspect. The numbers and types of ships at sea, their locations and their activities would be important information for Alliance leaders. While no one can judge in advance, or in any event with certainty, the meaning of naval movements for the calculus of deterrence, they give important indications of capability and intentions. Consider, for example, the spectrum of possible meanings indicated during a crisis by Pact maritime forces which:

- continue normal activity levels in port and at sea; or
- remain at sea with amphibious troops aboard beyond the estimated completion date of a large exercise and move away from the exercise area; or
- leave ports suddenly in large numbers and set course into the North Sea and the Norwegian Sea.

It is certain that in a crisis Western leaders would want continuous, detailed information about Eastern maritime activities; information which would be provided by increasing Western maritime and air surveillance — the various national technical means of surveillance alone could not provide all that would be wanted.

In addition to the West European naval and air forces which would be the first able to respond to the task, since few US forces are normally present in northern European waters, NATO has several forces in readiness and under command which might be dispatched to the Northern Region: the aforementioned STANAVFORLANT, STANAVFORCHAN, the AMF and the NAEW force. These units would contribute to deterrence not only by the information they gathered and their limited military capability but by their Alliance character. Whether and when any or all of their ships and planes might be found in Northern Waters in a crisis would, of course, depend on the overall situation, including the picture in other regions of the Alliance, and would be subject to the decisions of Allied governments acting through Permanent Representatives at NATO Headquarters in Brussels.

The defensive character of NATO is particularly apparent in its Northern Region. No Allied military units are based in Norway or Denmark. Iceland has no standing military forces, and those of Norway and Denmark are small. The region relies upon mobilisation and upon land and air reinforcement from Allies. The Rapid Reinforcement Plan (RRP) is thus particularly significant here.

Following approval of the RRP by NATO governments, the reinforcing air squadrons would fly in, but the arrival of troops reinforcing Norway, and to a significant extent the deployment of freshly mobilised Norwegian forces to northern Norway, requires safe passage through or near Northern Waters. The AMF is to an extent dependent on sealift if NATO's leaders call for its North Norway deployment option. The CAST Brigade, the UK/NL AF and the US 4 MAB all rely on sealift to reach Norway in full strength.

The same reliance on sea links is true of the Alliance as a whole. While Allied forces are based in some NATO nations, nowhere are standing forces present in sufficient strength to defeat a full Warsaw Pact attack without national mobilisation and reinforcement from North America and the British Isles. It would thus be an essential aspect of maintaining deterrence in a major crisis — for example, one in which the Pact had mobilised — for NATO to mobilise and reinforce. In such an event Northern Waters and air space would have a peripheral but significant role in the overall process.

The RRP relies initially on a vast, transatlantic air bridge to bring North American divisions to Europe. This string of the very largest military and civilian transport aircraft would require radar flight following for surveillance and rapidly available protection in the event of pre-emptive attack as they criss-crossed the Atlantic day and night for more than a week. Keeping watch on and above Northern Waters for movements of Soviet warships and long-range aircraft towards the air bridge would be an important precautionary measure by radar stations and aircraft from Norway, Iceland and the UK and perhaps calling also for carrier aircraft and seaborne radar.[26]

One common denominator of all that NATO must do to sustain deterrence in a major crisis is a prompt, sustained series of hard decisions. With respect to the Northern Flank, the military and political factors pose a daunting array of questions for national authorities and for NATO's North Atlantic Council, Defence Planning Committee and major NATO Commanders.

Major Soviet forces in the Kola are very much closer to Northern Waters and Norway's High North than equivalent forces of any NATO nation. How should Western actions take account of this fact? What would be the effect on deterrence if, before NATO's Striking Fleet Atlantic has been assembled or placed under SACLANT's command, the Northern and Baltic Fleets and aircraft deployed in numbers and in positions from which they could inflict severe losses on NATO forces entering Northern Waters if crisis turned into conflict?

Exercises, most recently SUMMEREX 85, have demonstrated a Soviet Fleet rapid deployment capability into the Norwegian Sea and south to the GIUK Gap and North Sea. During SUMMEREX and other exercises the Soviets have repeatedly demonstrated an impressive capacity to conduct large, co-ordinated attacks against ships by cruise missile-launching aircraft, submarines and warships. What does this augur for the reinforcement of Norway in a crisis, dependent as it is upon sealift?

The reinforcement of Norway — which prudence would seem to dictate should include capability to defend the reinforcements if attacked *en route* — would call into play maritime units from Canada, the Netherlands, the UK and the US. Is deterrence better served by conducting reinforcement operations under several national commands or NATO command?

It would be necessary to sustain confidence in the ability of Western forces to absorb a pre-emptive Pact attack if deterrence failed and still retain the ability to defend NATO's borders in the aftermath of that attack. In a crisis national and Alliance forces would require 'Rules of Engagement' which balanced the actions required to sustain deterrence while avoiding provocation, with the necessity to be confident that irreplaceable, mission-essential forces such as amphibious ships and troops and aircraft carriers would not be decimated by the tactical surprise which Pact forces might achieve in an initial blow. Northern Waters would be one focal point of this issue.

The US Navy's Maritime Strategy in Northern Waters reflects the obligation of the United States to defend the territory of all NATO members and the tasks and uncertainties involved in applying that commitment to the Northern Flank. It takes as a premiss that deterrence of war is the first goal. As we have seen, some of NATO's deterrent options require Alliance naval forces to operate in Northern Waters. The US Navy and Marine Corps will join that effort — indeed, finding European maritime forces already embarked on deterrent tasks in and near Northern Waters.

Admiral Watkins wrote that in a crisis Northern Waters would become the scene of increased patrol activity by US submarines and maritime patrol aircraft and that 'moving one Marine Amphibious Brigade to rendezvous with its prepositioned equipment and reinforce Norway provides a convincing signal of alliance solidarity'.[27] The CNO went on to say that

> Early forward deployment of sea based air power is also essential to support our allies (in deterring war), particularly Japan, Norway and Turkey. Early forward movement of carrier battle forces provides prudent positioning of our forces..., should war come. It does not imply some immediate 'charge of the Light Brigade' attack on the Kola Peninsula or any other specific target.[28]

The movement of NATO forces into the Northern Region in a crisis can produce a number of deterrent benefits: increased surveillance, through which Pact capabilities and intentions may be more accurately assessed by Western leaders. The physical presence of US and other ships, troops and aircraft not indigenous to the Northern Region would underscore the determination of NATO to protect all of its members. The early forward movement of maritime and air forces would make a dramatic difference to the defenders of northern Norway should deterrence fail: it would greatly increase the number of interceptors, add a significant SAM capability and provide aircraft for close air support. The presence of NATO anti-submarine warfare forces and carrier battle forces in or near Northern Waters would require the Soviet Union to hold a great deal of its maritime power in home waters rather than dispatch it to menace NATO's vital transatlantic sea (and air) lines of communication.

- **Horizontal escalation: an unlikely scenario**

During the Northern Waters Colloquium (see p. xvii), there were some assertions that the United States would respond to Soviet intervention in the Middle East by starting a major naval conflict in the Norwegian Sea. I believe that this view contains a misreading of several political and military factors which make it probable — I am tempted to say 'dictate' — that such a course of action is unlikely.

Soviet intervention in the Middle East would undoubtedly be of vital concern to all of NATO's nations; but unless it were the shared view of the Alliance that initiating a naval conflict in the Norwegian Sea was a necessary response, the United States doing so would irreparably damage NATO's political cohesion at a time when it had never been more necessary.

What would be gained by such horizontal escalation? It certainly would not put political pressure on the Soviets to withdraw their forces from the Middle East. It would be militarily dubious: devoid of NATO support, the United States would be granting the Soviets a huge advantage if it sought naval battles in the Norwegian Sea or mounted carrier strikes against the Kola military complex. (Indeed, if the United States were countering intervention by Soviet forces in the Middle East, there would not be much to spare for the Norwegian Sea.) A *guerre de course* would be no more promising than a series of fleet battles: to the extent that the Soviet Union has critical SLOCs, they run through the Black Sea and eastern Mediterranean via Suez to the Soviet Far East, not through the Norwegian Sea.

This is not a scenario to be considered seriously. Soviet intervention in the Middle East would raise enormous issues for NATO, not least of which would be the question of next moves by a Soviet leadership which was either so confident or so desperate as to take such a step; but US initiation of a naval conflict in Northern Waters, against the will of its Allies, would not be one of them.

- **In war: forward defence**

Should crisis in Europe become Warsaw Pact attack, NATO's intention is clear: defend all of the territory of the member nations, as far forward as possible, by conventional means if possible, while seeking to terminate hostilities and restore the *status quo ante*. Depending on the course of the preceding crisis, NATO's naval, air and land forces could be at full strength and in their intended defensive positions, or not. This would of course exercise a determining influence on the employment of maritime forces, in Northern Waters and elsewhere. These and other uncertainties are reasons that neither NATO's CONMAROPS nor the US Navy Maritime Strategy can be viewed as an operational master plan. Each is no more than a framework for approaching the problems of maritime deterrence and defence from the perspective of peacetime.

One aspect of maritime campaign planning which is clear to NATO's maritime commanders is that NATO does not dispose of enough maritime power relative to the Pact simultaneously to support the land battle throughout ACE and win another Battle of the Atlantic, unless the Pact remains on the defensive in at least some land and maritime theatres. The choices of where and when to bring maritime power to bear will be responsive to SACEUR's priorities, which are of course subject to the decisions of NATO political authorities. In that regard, a transition from crisis to war would bring no respite in the stream of hard choices facing the Alliance.

What might be the situation in Northern Waters if deterrence fails in the form of a Warsaw Pact attack on NATO? Since they are one path between NATO's Northern Flank and the Kola military complex and since they are one of the major avenues of Pact approach to NATO's essential transatlantic links, Northern Waters — like the North Sea, the Baltic and the Mediterranean — would undoubtedly be the scene of major engagements.

When? Where? What forces would be involved? What effects would they have on the scope and length of the war? Those questions cannot be answered in the abstract. Unsatisfying as it may be, in peacetime the most accurate answer to those questions is, 'It all depends'.

It all depends on variables such as:

- the placement of forces at the outbreak of hostilities;
- the situation in the aftermath of what Soviet writers have termed 'the battle of the first salvo';
- NATO's priorities for military action;
- the national decisions taken by NATO nations about the role of forces not under NATO military command; the role of the French Navy would be a significant factor;
- the degree to which the Pact respects Finnish and Swedish neutrality;
- the extent to which global Soviet military actions or vulnerabilities affect the forces which they and the United States commit to battle in the North Atlantic Treaty area.

Wartime NATO operations in Northern Waters would probably make two sorts of contribution to Alliance defence of the Northern Region. One would be supporting the battle ashore; another would be constraining Soviet maritime and air options to the south, which would occur as a result of not ceding either Northern Waters or Northern Flank territory by default. The Northern Flank campaign would be one of joint, coalition warfare in which for example, air force and marine aircraft from several nations operating from Norwegian airfields would be crucial to the efforts of multinational naval forces in Northern Waters. The efforts of naval forces, for example, in safeguarding logistic support or inserting and repositioning troops by amphibious means, would in turn be crucial to the defenders of North Norway.

- **Political control**

One aspect of discussion among conferees was concern that US naval activity on the Northern Flank, driven by the Maritime Strategy, would be of a unilateral, 'Rambo' character which might spin out of control and bring greater risk and harder fighting to the Northern Flank. Dangers on the Northern Flank and its potential involvement in military actions if deterrence fails stem from factors which exist quite independently of the US Navy Maritime Strategy:

- The growth of Soviet military power and its reach is an inescapable fact for the West. A most striking aspect of this growth is the military development of the Kola.
- NATO is an alliance based upon deterrence and, if deterrence fails, upon collective defence. This inevitably links events throughout NATO to the north and vice versa — indeed, it is the latter which underpins the national security of the Northern Region.
- It has never been NATO policy to cede the Norwegian Sea — or the Baltic or the Mediterranean, for that matter — or any of NATO's territory to an attacker.

It is these geopolitical facts of the 1980s which would bring fighting, if deterrence fails, to the Northern Flank. The US Navy Maritime Strategy is in part a reflection of these facts. It would be a mistake to confuse the strategy with the facts themselves.

NATO has a healthy, long-established political consultation process which maintains oversight and ultimately control of Alliance military strategy and operations. Political control of military action and of risk are at the heart of Alliance consultations on deterrence and defence. Given the voice of every NATO member in the consultation process and the certainty of compelling demands for naval forces elsewhere, it seems unlikely in the extreme that maritime forces would embark upon some unwanted enterprise in Northern Waters.

A specific focus of discussion about control of risk was the prospect of US attacks on Soviet SSBNs. This is a significant issue, as indeed is the refrain in Soviet military writing which indicates that they would make particular efforts in the initial stage of an attack on NATO to eliminate NATO's nuclear capability in Europe.

Anti-submarine warfare (ASW) is one of the most complex and crucial aspects of the defence of Europe. The Soviet Navy has more submarines, by a wide margin, than any navy in history. Those who remember what far fewer and far less capable German submarines did in the early years of the Second World War, and who recognise that sustaining NATO's defence depends on the safe completion of thousands of voyages, conclude that preventing a repetition of that near-strangulation is a NATO imperative.

The Maritime Strategy takes a broad approach to ASW:

- direct defence of the SLOC;
- an unremitting effort to delay, demoralise, damage or destroy each Soviet submarine, wherever found;

- the limitation of Soviet submarine operations against the SLOC by posing a threat which will cause Soviet authorities to withhold many submarines for other duties.

This US strategy proposes to make no attempt to exempt SSBNs from the general ASW campaign, and indeed seeks to tie down significant numbers of Soviet attack submarines and warships in the defence of SSBNs, which most observers believe would be seeking concealment in sea areas far north of NATO's SLOC.

There are compelling reasons for this approach. One is the obvious benefit to NATO if the Soviets rally attack submarines and powerful surface vessels to the defence of their SSBNs, because these same forces cannot then simultaneously threaten NATO's SLOC. Another is that since the Soviet SSBN force is a significant strategic asset, a steady loss of SSBNs, if accompanied by a discouraging outlook for victory on the ground in Europe, would be a factor in convincing the Soviet leadership that they should terminate hostilities. A third reason for this approach is the practical consideration that in ASW it is important to attack as rapidly as possible after one is confident that a hostile submarine has been located. Not only is there the prospect that the quarry will realise the danger and escape, a submarine is a deadly opponent for another submarine or a surface ship, one which can quickly turn the tables and send the hunter to the bottom. Time spent in sorting out which type of Soviet nuclear submarine has been located can be time in which the engagement is lost.

NATO political authorities could be expected carefully to assess whether continued losses of SSBNs would lead the Soviets to escalate a conventional attack on Europe to nuclear war, for example, because it placed them in a 'use them or lose them' dilemma. While one cannot predict the course of NATO consultations on this and other aspects of NATO's Flexible Response strategy, current US Navy thinking is that one could reasonably contemplate the attrition of the Soviet SSBN force during an unrestricted ASW campaign because it would not create the use or lose dilemma, which requires aspects of speed and certainty not present in ASW operations.

The destruction of the large Soviet SSBN force would not occur with great speed because the uncertainties of ASW prevent the simultaneous destruction of a large number of deployed SSBNs. This ensures that the time available for dealing with attrition of a large SSBN force would be days and weeks, not minutes, as could be the case in a threat to fixed, land-based missiles. There is no possibility of the Soviet leadership being placed in a situation from which they have only minutes or hours to reach a decision on how to respond.

The second point which obviates the occurrence of this cataclysmic dilemma is that there would be a number of Soviet remedies to the withering away of their SSBN force. Thus, for example, they could either strengthen forces defending SSBNs or abandon a failed 'bastion' strategy and disperse their SSBNs and protecting forces into the Atlantic. Dispersal over a wider area would greatly increase the survivability of SSBNs over that existing in Northern Waters if bastion defences had proven to be ineffective. One can envision a number of options involving 'hotline' communication between Soviet and American leaders by means of which the Soviets could seek to salvage their SSBN situation short of a Soviet launch order.

In sum, the possibility of Soviet escalation to nuclear warfare in response to attacks upon their SSBN force cannot be ignored (and the US Maritime Strategy does not ignore it), but a global ASW campaign is a relatively slow process which would not place the Soviets in a dilemma of conclusive and fatal choices. It would, however, increase the cost to Soviet power of continuing to wage war against NATO, which is one of the reasons for a policy which contemplates such attacks.

- **A new context creates new requirements**

The Soviet build-up in the Kola over about the past two decades has fundamentally altered the political–military context in which Northern Waters are set. The year 1988 differs entirely from 1968 because if the Soviets attacked today, a major NATO effort would be necessary in Northern Waters to defend against Soviet maritime power. Their 'fleet in being' would need to be destroyed or forced to remain on the defensive in positions where it could not help turn the Northern Flank or cut NATO's SLOC.

Furthermore, and perhaps of equal concern, Soviet maritime strength relative to that of NATO gives them options for intimidation short of war which they did not have in 1968. We should keep in mind the danger described in the 1975 British White Paper on Defence, that 'If the balance of maritime power were to shift so far in favour of the Warsaw Pact that it had an evident ability...to isolate Europe by sea, the effect on Allied confidence and cohesion would be profound'.

The Soviet build-up in the High North puts considerable new pressure on the Northern Region and on Nordic security policies, which strive to maintain a balance of reassurance and deterrence towards the Soviet Union. At the same time, it alters the context of those policies. In the new context, maritime activity in Northern Waters, which was once important primarily to the defence of the Northern Flank, is now also critical to sustaining the defence of the Central Front. Furthermore, in the new context of Soviet maritime power based on the Kola and demonstrably ready to deploy rapidly and competently throughout Northern Waters and beyond, Western military activities in the Northern Flank which could have disturbed the balance in 1968 would not do so in 1988.

The deterrent side of the balance needs further attention, first of all by the Nordic nations themselves who, to be candid, sometimes give Americans the impression that their primary concern is reassurance of the Soviet Union. The description of the situation in the late 1970s, quoted previously in this chapter, still, in my view, applies:

> The Soviet naval buildup....tended to cast doubts on the assumption that allied assistance could be brought to bear on a crisis... Soviet submarines and naval aircraft with Standoff weapons emerged as a potent threat against its real effectuation in an emergency.[29]

The emphasis in the US Navy Maritime Strategy on early forward movement of major naval forces in a crisis is a clear statement of intent to bring Allied assistance to bear despite that threat. To those who understand that the alternative to these deterrent actions is Soviet domination of Northern Waters, this strategy should be encouraging news.

In attending to deterrence, not only must the pre-positioning decisions already taken be carried out fully, more needs to be done in both political and military terms to facilitate the timely arrival of adequate NATO forces in a crisis. This work will increase the odds that Northern Waters will continue to be only an area of peacetime operations.

NOTES

1. Iceland is omitted from this list since although it has maritime interests and activities it has no national military forces. The term 'home waters' is used here to denote those areas including but extending well beyond the territorial sea where a nation's maritime forces routinely operate.

2. The Agreement on the Prevention of Incidents On and Over the High Seas dates from 1972. It consists of special signals which may be used by ships and aircraft to communicate intentions and thus avoid ambiguous or potentially

dangerous situations. It includes undertakings to refrain from specific actions which could appear threatening or could cause a hazardous situation. The convention is reviewed formally at regular intervals by the two navies meeting at the Flag Officer level alternately in Washington and Moscow.

3. Johan Jorgen Holst, 'The security pattern in northern Europe: a Norwegian view', paper delivered at a conference on 'Britain and the security of NATO's Northern Flank' arranged by the Department of War Studies of King's College London, 7–8 May 1986, p. 8.

4. Prime Minister Kaare Willoch, 'Norway's security policy situation', speech to the Norwegian Military Society, Oslo, 7 April 1986.

5. Johan Jorgen Holst, 'Norway's search for a Nordpolitik', *Foreign Affairs*, Fall 1981, p. 72.

6. Ministry of Defence, Federal Republic of Germany, *The German contribution to the common defence* (Ministry of Defence, Bonn, 1986), p. 12.

7. Vice Admiral Helmut Kampe, GeN, 'Defending the Baltic approaches', *US Naval Institute Proceedings*, March 1986, p. 93.

8. Colonel Jonathan Alford, Deputy Director, International Institute for Strategic Studies, paper delivered to the conference on 'Britain and the security of NATO's Northern Flank'.

9. Quoted by Dr Geoffrey Till in his paper 'British naval perspectives on the far north', delivered to ibid.

10. Although CONMAROPS itself is a classified document, the essential features which will be described in this chapter are in the public domain. See for example 'NATO antisubmarine warfare', *North Atlantic Assembly Papers*, 1982; Admiral W. McDonald, US Navy, 'Mine warfare: a pillar of Maritime Strategy', *US Naval Institute Proceedings*, October 1985; Vice Admiral H. Mustin, US Navy, 'The role of the Navy and Marines in the Norwegian Sea', *Naval War College Review*, March–April 1986.

11. On this matter, both the previous (Conservative) government and the present (Labour) government seem to agree. Prime Minister Willoch in his 1986 speech to the Military Society and Defence Minister Holst in a Reuters interview given 1 July 1986, each called for a return to previous levels of Allied naval activity in the Norwegian Sea. This is, fortunately for a thinly stretched US Navy, not a call for a permanent presence there. As Holst told Reuters, 'It is not in anybody's interest to make it into another Mediterranean'.

12. Norman Friedman, 'US maritime strategy', *International Defense Review*, no. 7 (1985), p. 1070.

13. The authoritative public exposition of the US Navy Maritime Strategy is found in three articles published as a supplement to the January 1986 issue of *US Naval Institute Proceedings*. The supplement contains articles by the Secretary of the Navy, the Chief of Naval Operations and the Commandant of the Marine Corps plus a useful bibliographic essay. Reprints may be obtained from the Editor, *US Naval Institute Proceedings*, Annapolis, MD 21402. See also Rear Admiral William Pendley's response to questions about the foregoing articles in the June 1986 *Proceedings*, pp. 84–9.

14. This two-paragraph overview of US defence strategy is drawn from Secretary of Defense Caspar W. Weinberger's annual report to the Congress for FY 1987, Executive Summary pp. 6–12.

15. General Bernard W. Rogers, Supreme Allied Commander Europe, 'NATO strategy: time for a change?', *Alliance Paper*, no. 9, (The Atlantic

Council of the United States in Co-operation with the US Mission to NATO, Brussels, October 1985), p. 5. The General concluded that 'despite all the critical attention our current strategy receives, there are no convincing arguments — at least in my mind — that it should be discarded'.

16. The NATO Information Service, *NATO facts and figures*, (NATO, Brussels), p. 153.
17. Ibid., p. 144.
18. McDonald, 'Mine warfare', p. 47.
19. Ibid., p. 48.
20. Mustin, 'The role of the Navy and Marines', p. 2. Admiral Mustin was then the Commander of NATO's Striking Fleet Atlantic.
21. Special supplement, 'The Maritime Strategy', *US Naval Institute Proceedings*, January 1986.
22. McDonald, 'Mine warfare'.
23. Special supp., 'The Maritime Strategy', p.8.
24. Ibid., p. 5.
25. Ibid., p. 11.
26. While the NATO transport aircraft would be at high altitudes, near the edge of Soviet naval SAM capabilities and Soviet aircraft with range to reach the air corridors are bombers rather than interceptors, both warships and bombers could pose a threat to the air bridge which NATO would wish to counter. The huge, relatively clumsy transports are enormous targets in comparison with those against which naval SAMs were designed, so that even at the edges of the SAM capability envelope the danger should not be disregarded. Bombers might be jury-rigged to launch heat seeking-missiles which in the circumstances could be effective even if the aiming and launching devices were crude. (The author is indebted to Admiral 'Ike' Kidd, US Navy Retired, formerly SACLANT, for this point.)
27. Special supp., 'The Maritime Strategy', pp. 9–10 and 26. When prepositioning of equipment for 4 MAB has been completed in 1989, this deployment could be via aircraft. Until then, 4 MAB would deploy by ship.
28. Ibid., p. 10.
29. This is drawn from Johan Jorgen Holst's 1986 paper (as noted). The full statement is quoted on p. 181 of this chapter.

10

The Maritime Strategy and Geopolitics in the High North

Steven Miller

For the last several years there has been growing discussion in the United States on the evolution of US naval strategy. In the early 1980s the US Navy undertook a re-evaluation of its global strategy, the result of which now goes by the name 'The Maritime Strategy'. Elements of this strategy have been in public view in the United States since at least 1981 when the Reagan Administration came to power and signalled its intention to strike out in new directions in American defence policy. By early 1986 the US Navy had fully embraced the Maritime Strategy and had coherently and publicly described and advocated the strategy in all its dimensions. In response, there has emerged in the United States a heated controversy over the appropriateness and desirability of the Maritime Strategy.

Europe has been slow to recognise this important development in US policy, although it has significant implications for NATO and, particularly, for Norway. For three reasons it is understandable that this is so. Firstly, while aspects of the new strategy were evident in the American debate, nevertheless for some time the full and exact nature of the strategy was not clear to the public eye (and further, the strategy seems to have evolved over time), with the result that there was considerable uncertainty and ambiguity about what the Maritime Strategy was and what its implications might be.[1] Secondly, this development was the product of internal deliberations within the US Navy, not the result of discussion in Brussels or in other fora in which the European Allies participate. Thirdly, many of the public indications about the Maritime Strategy were scattered untidily in a variety of sources — Senate and Congressional hearings, professional journals, specialised defence publications, and so on — that are not easily accessible to those who do not follow the American naval debate at a detailed level. Consequently, discussion in Europe of the Maritime

Strategy has lagged a year or two behind that in the United States: the reality of the Strategy has not been fully accepted, its implications have not been fully appreciated, and its importance has not been fully realised. This seems true even in Norway, the NATO country most directly affected by the American adoption of the Maritime Strategy.

THE EVOLUTION OF THE AMERICAN DEBATE

In the United States, the deliberations that culminated in the Maritime Strategy are dated at least to 1979, when senior leaders of the US Navy, concerned by the naval policies of the Carter Administration, sought to provoke discussion on the future of the Navy.[2] The coming to power of the Reagan Administration in 1981 galvanised this process of rethinking, as the Administration early on committed itself to a substantial naval build-up and to a 'rebirth' of US naval strategy. The new Secretary of the Navy, John F. Lehman, Jr, wrote shortly after taking office:[3]

> The US Navy will grow in size and capability into a fleet which possesses outright maritime superiority....Nothing below clear superiority will suffice....The US Navy of the near future will be visibly offensive in orientation....It will stress forward deployment, including operations capable of war-fighting and winning in areas denoted as 'high risk'....The broader equipage of the US Marines, and the marked growth of amphibious sealift, will enhance our capability to wrest important coastal land areas and chokepoints to our control when necessary.

On other occasions, Secretary Lehman was even more explicit about the strategy he had in mind for the Navy. As *Newsweek* reported in March 1981,[4]

> In a breakfast meeting with reporters last week, Navy Secretary John Lehman announced *a bold new strategy* for the US Navy. With fifteen mighty carrier task forces, the Navy would gird itself to sail directly into harm's way and take on Soviet forces just off the Russian Coast.

This would involve 'steaming into Soviet strongholds such as the Kola Peninsula near Finland and the Kamchatka Peninsula near Japan to bottle up the Russian Navy'.

Such commentary revealed, even five years ago, the central precepts of the Maritime Strategy in broad outline. In the intervening years, these broad instincts have been translated and refined into a clear, coherent, and comprehensive strategy, one that contains an explicit concept of how the war at sea may unfold, what kind of sequence of operations would be required of the US Navy, and how those operations would contribute to the winning of the war.[5]

The 'Maritime Strategy' hearings held in March 1984 by the Subcommittee on Sea Power and Power Projection of the Senate Armed Services Committee represented something of a turning point in the public discussion of US naval strategy, for with their publication most of the key elements of the Maritime Strategy were now authoritatively in the public view.[6] The issues and controversies associated with the Maritime Strategy were increasingly, openly, and vigorously under discussion.[7]

Two issues, which have remained central to the debate over the Maritime Strategy, have been particularly controversial. Firstly, how much, if anything, does the Strategy contribute to the deterrence of war in Europe? Secondly, how feasible and, more importantly, how escalatory are attacks on Soviet strategic submarines?[8]

Some answers to these two questions — and others — can be found in the substantial defence of the Navy's position published in September 1985 in *United States Naval Institute Proceedings* (*USNIP*) and in what has been called the Maritime Strategy Whitepaper, a supplement to the January 1986 issue of *USNIP* which contained essays by Chief of Naval Operations, Admiral James Watkins, the Commandant of the Marine Corps, General P.X. Kelley, and the Secretary of the Navy, John Lehman. With the publication of this document, there was now an authoritative source for the Maritime Strategy and there could no longer be any doubt that important new elements had been introduced into the debate on US naval strategy.[9] This was *not*, as it may seem to many in Europe, the beginning of the public debate, but rather a significant progression in an ongoing debate. However obscure much of this may seem to America's NATO allies, in the US the key issues and controversies have been in the public eye for some time.

IS THE MARITIME STRATEGY REAL?

But how real is all this, how seriously should the Maritime Strategy be taken? Many, particularly in Europe, seem to believe (or to hope) that there is less to the Strategy than meets the eye. One variant of this belief suggests that it represents the idiosyncratic preferences of an individual (namely, Secretary of Navy, Lehman) or a small group (namely, Lehman in league with Admiral Watkins and a few other senior naval officers) who are isolated from the day-to-day operational realities of the Navy. From the Navy's perspective, however, the Maritime Strategy represents a great renaissance of strategic thinking, one which pervades the entire service. As Captain Linton Brooks explains,[10]

> the strategy embodies the professional consensus of the leadership of the Navy and Marine Corps on how to deter or, if necessary, fight, a future war... The strategy provides a common frame of reference for Navy and Marine Corps officers... A significant effort is underway within the Navy to ensure its officers understand the strategy and their role in it and are active in its continued refinement.

A second variant of the belief that the Maritime Strategy is less important than it seems suggests that the strategy is primarily budget rhetoric, that its main purpose is not to govern the thinking, procurement, and activities of the Fleet but to persuade Congress and the public to support the Reagan Administration's naval build-up. While it is of course true that the Strategy provides a strong rationale for heavy investment in naval capability, to conclude that it is nothing but a scheme to enhance the Navy's budget is to ignore the discernible impact it has had and is having in a number of aspects of naval policy. Perhaps the most visible evidence of this is the increased emphasis on US and NATO naval exercises in the Norwegian Sea. In the autumn of 1985, for example, the exercise Ocean Safari included the deployment of a carrier battle group in the Norwegian Sea.[11] Vice Admiral Henry Mustin, Commander of NATO's Atlantic Strike Fleet, has called for at least a doubling of the amount of time that US carriers spend in the Norwegian Sea — which does not amount to a huge increase because in existing deployment and exercise patterns they spent very little time in the Norwegian Sea, but which does reflect the growing commitment to forward operations.[12]

This is not the only impact of the Maritime Strategy on the Fleet. It has also influenced logistical arrangements, necessitated by the anticipation of heavier operations in forward areas, and Fleet organisation (which has been altered to make the Fleet more effective at strike missions).[13] Further, the Strategy is helping to shape the Navy's procurement programme, a good example being the development of the SSN-21, a new attack submarine intended to have improved under-ice capability so that it can effectively track Soviet strategic submarines in the Arctic. As Admiral Watkins put it, 'The strategy has become a key element in shaping Navy programmatic decisions... The Maritime Strategy has rationalized, disciplined, and focused Navy program development, budgets, and procurement to a degree that would have seemed remarkable five years ago.'[14] Clearly, the Strategy is more than just rhetoric.

A third argument of those who would dismiss or ignore the Maritime Strategy claims that because the strategy did not emerge from the NATO bureaucracy in Brussels it is not official NATO policy and therefore cannot be real. To believe thus is completely to misunderstand the perspective of the US Navy. It views the Strategy not as a rejection or modification of US commitments to NATO, but as a fulfilment of them. Indeed, the Strategy is made imperative by America's allies. Alternatives to the Maritime Strategy, Admiral Watkins has stated, 'would inevitably lead to the abandoning of our allies... Allied strategy *must* be prepared to fight in forward areas'.[15] Similarly, Secretary of the Navy, Lehman rejects criticisms of the Maritime Strategy with the observation that

> No coalition of free nations can survive a strategy which begins by sacrificing its more exposed allies to a dubious military expediency. To suggest that naval support of Norway or Turkey is too dangerous because it must be done close to the Soviet Union is defeatist.[16]

In addition, the naval contribution of America's NATO allies is counted on to help make the Strategy feasible. As Captain Brooks explained, 'Throughout all phases of the Strategy, close cooperation with allied navies... is mandatory.'[17] American involvement in NATO, in short, has been an integral component of the case for the Strategy.

Clearly, it is a mistake to accept the arguments that suggest that the Maritime Strategy should not be taken seriously. There is every indication that it represents the genuine and deeply held strategic conclusions of the US Navy; to ignore it is to miss a significant development in the evolution of US security policy.

WHAT IS THE MARITIME STRATEGY?

If the Maritime Strategy must be taken seriously, then its purpose and contents become matters of great importance.[18] The Strategy has not one, but several separate purposes. It is intended to provide vigorous support of allies, including Norway, Japan, and even Turkey. A second purpose is to defend NATO's sea lines of communication (SLOCs), which would become vital to the provision of reinforcements should the war not end quickly. Thirdly, naval power will be projected against Soviet targets ashore in order to create confusion and uncertainty in the Soviet rear area and to tie down or perhaps even divert Soviet forces that might otherwise be employed in the battle in Central Europe. A fourth purpose is to threaten the Soviets with adverse changes in the nuclear correlation of forces, thereby enhancing deterrence and providing war termination leverage should war come. Finally, the Maritime Strategy is intended to thwart the preferred Soviet vision of the war by ensuring that that war will not be short and focused on Europe, as the Soviets appear to desire, but rather protracted and global in scope. Taken together, this collection of purposes is thought to provide a very effective deterrent strategy; obviously, if one can successfully defend Norway, protect NATO's SLOCs, attack Soviet territory by air, force the Soviet Union to capitulate by destroying their strategic submarine fleet, and cause the USSR to fight a war it does not want to fight, then it is possible to offer a powerful deterrent threat. Additionally, however, the Navy believes that this set of ideas represents an effective warfighting strategy should deterrence fail. It is certainly a worthy objective to seek an effective deterrent and warfighting strategy, and these goals in themselves are not, for the most part, especially controversial.

These objectives are, of course, very ambitious (and hence provide a broad and important rationale for naval power in NATO's defence posture). There are two very significant observations that derive from the ambitious scope of the Maritime Strategy's multiple objectives. Firstly, these missions do not all lead to the same operational requirements. Thus, for example, operations very far forward (in the Barents Sea) may be inescapably necessary to the campaign against Soviet strategic submarines or to amphibious operations on the Kola Peninsula, but unnecessary for the defence of Norway or the protection of the North Atlantic SLOCs. Keeping the separate purposes of the Maritime Strategy distinct is therefore important in assessing thoroughly the feasibility and desirability of its various components: there is no reason to suppose that all elements of the strategy will be equally viable or attractive. Secondly, there may be important tradeoffs among these objectives in terms of timing and priorities. While some operations can support multiple objectives, it will not always be the case that forces committed to one have the utility in pursuing others. Attack submarines committed to chasing Soviet strategic submarines in the Barents Sea or Arctic Ocean, for example, will not be available for operations south of Iceland should the USSR decide to mount a serious attack against NATO's North Atlantic SLOCs.[19] Aircraft carriers committed to the support of US Marines in Trondheim (in south-central Norway) or to the defence of Norway along the Bodo-Tromso line will not be free to steam into waters further north.[20] The Strategy does not call for the simultaneous performance of its various missions, but critical questions of priority and sequence could well play a major role in determining its ultimate impact on the fortunes of the war — and on the fate of Norway.

The Maritime Strategy provides not only a set of objectives for the US Navy, but also a vision of how the war at sea will unfold and how the Navy would achieve its objectives. It conceives of a war in three phases. In the first phase, the transition to war, the essential element of the Strategy is the massive global forward movement of naval forces. The idea, obviously, is to have as much capability as possible — particularly attack submarines — in place in forward waters and ready to fight, should the war begin. This puts a premium on early recognition of warning and prompt, timely decision-making by political and military authorities. As Admiral Watkins has stated, 'Keys to the success of both the initial phase and the strategy as a whole are speed and decisiveness in national decisionmaking'.[21] The greater the delay, during a crisis, in sending US and NATO naval forces forward, the longer and more difficult the naval campaign will be.

In the second phase of the war as conceived by the Maritime Strategy, the US Navy will attempt to 'seize the initiative' by establishing sea control as far forward as possible. The hallmark of this phase will be two campaigns of attrition against Soviet forces. The primary effort will be an anti-submarine warfare (ASW) campaign against Soviet submarines, both tactical and strategic. The destruction of Soviet attack submarines will break open the Soviet bastion, reduce the most serious threat to US forces at sea, permit freer and more aggressive US naval operations, while also increasing the vulnerability of Soviet strategic submarines whose protection is a major (if not the major) job of Soviet attack submarines. The destruction of Soviet submarines will cause a shift in the nuclear correlation of forces against the USSR and thereby, so the US Navy believes, put mounting pressure on the Soviet leadership to terminate the war by suing for terms. While the ASW battle is the centrepiece of phase II of the Maritime Strategy, the battle in the sky to impose attrition on Soviet airpower in the northern European theatre is also very important. Soviet air power (and particularly the Backfire bomber armed with air-to-surface missiles) poses a serious threat to NATO bases and forces in Norway and in the Norwegian Sea. Weakening the Soviet air threat will reduce the vulnerability of crucial Norwegian bases and will permit American aircraft carriers to operate in greater safety. In addition to ASW and anti-air operations, efforts will also be made in the second phase of the Strategy to destroy Soviet surface ships.

Success in phase II of the Maritime Strategy will liberate the US Navy to conduct more aggressive operations in phase III, 'carrying the fight to the enemy'. As Admiral Watkins has explained,

> Successes in anti-air, anti-submarine, and anti-surface warfare are crucial to effective prosecution of offensive strike warfare... To apply our immense strike capability, we must move carriers into positions where... they can bring to bear the added strength needed on NATO's Northern or Southern Flanks, or in Northeast Asia.[22]

Against a significantly weakened, if not shattered, Soviet air and sea threat, US carrier battle groups will be able to operate with greater confidence and aggressiveness further forward in support of land attack and amphibious operations against Soviet territory. Similarly, US attack submarines will be free to prosecute their campaign against Soviet strategic submarines with great vigour. At this point, sea control will have been established and the aims of continued operations will be to complete the destruction of Soviet naval power, to contribute to the battle in Central Europe by putting serious pressure on the Soviet flanks, and to attempt to induce favourable war termination by destroying the Soviet strategic submarine force.

It is important to recognise that there is no timetable attached to the several phases of the strategy, each may be as long or as short as is dictated by prevailing conditions. As noted above, for example, hesitation in moving forces forward in phase I will likely cause phase II to be more protracted and arduous. Similarly, phase III operations will depend on some minimum level of success in phase II. Attempting phase III operations prematurely, before sufficient attrition had been exacted against Soviet forces, would expose US carrier battle groups to excessive risks. Conversely, should US forces meet with unexpectedly rapid success in phase II, the transition to phase III operations could come more quickly.

It is also important to recognise that there is no clear-cut line between phase II and phase III operations: the differences are more ones of degree than of kind. It is probably more accurate to conceive of the strategy as a series of sequential operations, the earliest of which place more emphasis on submarine operations, the later of which highlight forward carrier operations. At every step, however, both carriers and submarines (as well as land-based air power) will have some role to play. The US Navy has been emphatic that the Maritime Strategy does not call for a mad, reckless dash far north with carriers into the full brunt of Soviet capability: critics correctly point out that this would not likely be profitable, but this observation does not undermine the Strategy because such action is not contemplated. While forward operations will never be without risk, the US Navy seems convinced that the phased operations called for by the Strategy will not place US forces in positions of extreme jeopardy.[23]

Such, in brief, is the concept of naval operations in wartime that is provided by the Maritime Strategy. It is clearly a very ambitious and an offensive strategy, but one that the US Navy believes is feasible, will contribute to deterrence, and will achieve essential objectives if war comes.

HOW NEW IS THE MARITIME STRATEGY?

Does the Maritime Strategy represent a dramatic change in US naval strategy? Sometimes it is represented as a bold new shift in direction, a 'strategic renaissance'; at other times, the emphasis is on the elements of continuity between the Strategy and earlier US naval policy. This is a difficult question to disentangle, in part because the earlier operational preferences and expectations of the Navy are not clearly visible to public view, in part because (as is often the case in such matters) there are elements of both continuity and divergence in the Strategy.

It is certainly *not* new for the US Navy to prefer, and to have, an offensive strategy. At least since the time of the great US naval strategist Alfred Thayer Mahan in the late nineteenth century, the US Navy has had a strong preference for offensive doctrines built around the concept of the decisive battle: the proper function of the battle fleet, in this view, is to seek out and to destroy the naval power of the adversary.[24] The Maritime Strategy is completely consistent with this tradition.

It is also *not* new for US naval doctrine to include forward operations against Soviet naval capabilities.[25] This is a natural concomitant of offensive naval strategies. Furthermore, the US Navy has long believed that there are great advantages to operating its forces — especially attack submarines — in forward areas: it puts the Soviets on the defensive, enhances Western ASW capabilities, increases the difficulty of Soviet transit to the North Atlantic, and so on. Throughout the 1970s the US Navy gave clear indications of its commitment to forward operations.[26]

It is also likely that plans to attack Soviet strategic submarines are *not* new. The more that becomes publicly known about American nuclear war plans, the more clear it is that attacks against Soviet nuclear forces have been given high (if not highest) priority.[27] It would be very surprising if the US made serious efforts, sustained over several decades, to threaten Soviet land-based nuclear capability, but left its sea-based forces unthreatened. Hence, it is reasonable to assume, in the context of a nuclear war, that the US Navy has always intended to destroy Soviet SSBNs as best it can.

There are, in short, considerable elements of continuity between these traditional doctrinal instincts of the US Navy and the current Maritime Strategy. The extent to which the latter represents a dramatic change in direction should not be overstated. Nevertheless, there do appear to be some aspects of the Strategy that are new (although it is necessary to be at least somewhat tentative on some points since we cannot be confident that the public record provides a full picture of earlier US naval strategy).

At a minimum, it is new that these matters are being articulated so openly and debated so publicly. The Navy is often said to be the most secretive of the American military services and it is quite unusual to have such a comprehensive and detailed discussion of naval strategy and operations. Furthermore, it is virtually unprecedented to have available a thorough and coherent description and explanation of US global naval strategy such as is provided by the Maritime Strategy. From a *political* point of view, these may be the most salient facts: the relevant publics in the US and NATO Europe will be confronted with many of these issues for the first time, regardless of how new the Strategy actually is.

From a *military* perspective, however, there are several dimensions of the Maritime Strategy that may be new, or at least that seem to be reserving greater emphasis than in the past. One is the great premium placed on large-scale early forward movement of forces. Another is the focus on purposely seeking out and attacking Soviet strategic submarines from the *outset* of a *conventional* war. A third is the priority given to threatening Soviet territory with air and amphibious attack. As near as one can tell, these seem to constitute new elements introduced by the Strategy.

Although there has been interest in the question of how new the Maritime Strategy really is, the far more important point is that the virtues and disadvantages of the various elements of the Strategy are unchanged regardless of whether it is an example of continuity or innovation: its advantages are no more attractive if it is new; its disadvantages are no less unattractive or dangerous because it is old. What is necessary is to examine the components of the Strategy on their merits, to assess their costs and benefits independent of the extent to which they represent strategic innovation.

CRITICISMS OF THE MARITIME STRATEGY

This means, of course, that we must consider the case against, as well as that for, the Maritime Strategy. This Strategy has been quite controversial in the United States. As former Assistant Secretary of Defense Frances West has written (perhaps with a little exaggeration), 'There are about ten articles written attacking the strategy for every one written to explain it.'[28] Whatever the true proportion of critics to advocates, it is certainly the case that there have emerged several significant families of criticism of the Strategy.[29]

The criticisms of the Maritime Strategy basically derive from four fundamental concerns about the operations it requires:

(1) Are they feasible?
(2) Will they be effective in providing advertised benefits?
(3) Are they dangerously risky or escalatory?
(4) Are there preferable alternatives?

In what follows, I will sketch out briefly the nature of these concerns.

(1) Feasibility

The forward naval operations envisioned by the Maritime Strategy require that the US Navy (along with Allied naval support) confront the Soviet Navy at its strongest points, in areas (such as in or near the Barents Sea in northern Europe and the Sea of Okhotsk in the Pacific) where the USSR can sustain large concentrations of forces under cover of land-based naval aviation. In such areas, crucial US advantages in naval capability — its superb logistics support in distant waters, its carrier-based air support on the high seas, etc. — are negated: when the Soviet Navy can operate close to home, its logistics problems are much smaller and its air cover from bases ashore is quite potent. The Maritime Strategy, in short, calls for operations in precisely those few ocean areas on the planet where the USSR can, or at least may be able to, give the US Navy no less than an equal fight.

It is for reasons such as these that some critics question whether the US Navy possesses (or will in the future possess) sufficient capability successfully to implement the Maritime Strategy. Some believe that the US does not possess enough attack submarines, armed with appropriate weaponry, effectively to implement the campaign against Soviet SSBNs; and it is also suggested by some that more generally, improvements in submarine technology have outpaced advances in ASW capability.[30] As one of the strongest critics of the strategy, John Mearsheimer, concluded, 'No evidence in the public record would lead one to view counterforce coercion as a high-confidence option'.[31]

Even more doubt has been raised about the wisdom of operating aircraft carriers forward and thereby exposing them to the maximum threat; their vulnerability in these regions, critics fear, is simply too great for these operations to be worthwhile. Defence analyst Jeffrey Record has written, for example, that[32]

> so aggressive a naval doctrine requires bringing the carriers into waters infested with Soviet submarines and well within range of Soviet air power dedicated to the destruction of the US surface navy... To venture US carrier battle groups close enough to the Soviet Union to launch air strikes on the Soviet Navy's home ports is to venture into the jaws of defeat.

Another prominent doubter is retired Admiral Stansfield Turner who, in an essay written with Captain George Thibault, argued that forward carrier operations could not be undertaken without suffering serious losses and concluded, in a much-quoted passage,[33]

> The loss of three or four of the Navy's 12 to 13 carriers, in what would have to be a gamble to suppress the Soviet Navy in this manner, would be a major catastrophe. No President could possibly permit the Navy to attempt such a high risk effort.

William Kaufmann believes that current and planned US naval forces fall short of those necessary to conduct forward carrier operations with any confidence. He has written that[34]

> if the attack depended only on the power of three carrier battle groups, the odds are rather high that all three would be disabled or sunk and that a large part of the Soviet fleet would survive. Indeed, if the Navy is determined to sail into harm's way with surface combatants in the Norwegian and Barents Seas, it had better to do so with at least nine carrier battle groups...

...leading to an overall requirement for 24 carriers!

The US Navy, despite some debate within about tactics and timing and despite some internal dissent, of course rejects these criticisms: it would not advocate a strategy it did not believe it could implement. As Admiral Watkins wrote, the Maritime Strategy 'is a strategy for today's forces, today's capabilities, and today's threat'.[35] Commenting on the US ability to destroy Soviet submarines in Northern Waters, Watkins commented in 1984 that 'we know how to get inside there and find them, trail them, and kill them'.[36] Admiral Mustin, Commander of the Atlantic Strike Fleet, has been particularly emphatic on the point that Soviet forces, though formidable, will not be invincible, and concludes that NATO naval forces 'are eminently capable of carrying out our strategy successfully'.[37] But although the Navy's senior leadership expresses confidence, the critics remain unpersuaded and doubts continue to be voiced.[38]

(2) Effectiveness

A more damaging line of criticism suggests that even if the Maritime Strategy is feasible and is successfully implemented, it will not provide the benefits promised by the Navy. Two elements of the strategy in particular have been questioned on these grounds: aerial and amphibious attacks on the Soviet flanks for the purpose of pinning down and/or diverting Soviet forces that might otherwise be used on the Central Front; and attacks against Soviet SSBNs for the purpose of inducing the Soviets to seek war termination. Neither of these operations are seen as having the effect that the Navy claims.

With regard to attacks on the flanks, critics argue that the Soviets will not allow such operations to interfere with their campaign in the primary theatre in Central Europe, and furthermore suggest that the USSR is not greatly vulnerable to the exercise of naval power. As one of the Maritime Strategy's most severe critics, Robert Komer, argues in an influential article in 1982, 'Sweeping up the Soviet navy, nibbling at the USSR's maritime flanks... would hardly suffice to prevent a great Eurasian heartland power from dominating our chief allies...'[39] Edward Luttwak criticises the strategy in similar terms: 'The Soviet Union... could invade Western Europe, reduce China, and neutralize Japan without sending a single ship to sea; its war economy could survive indefinitely without any maritime communication; and its own territory is shielded from the oceans...'[40] Clearly, in the view of such critics, maritime attacks on the Soviet flank are going to provide little benefit.

The effectiveness of attacks on Soviet strategic submarines is likewise questioned. The US Navy claims that this campaign will bring about 'war termination leverage' against the USSR — that is, that it can force the Soviet Union to surrender in some fashion, to sue for peace on unfavourable terms. This will come about, the Navy believes, because the 'nuclear correlation of forces', about which the Soviets seem to care deeply, will be moving against the USSR. However, critics point out that even if every Soviet strategic submarine were destroyed at sea, Soviet land-based strategic forces (notably its ICBM force), the heart of Soviet nuclear capability, would remain untouched and the USSR would therefore still have in its possession thousands of highly capable nuclear warheads. Under such circumstances, it is not obvious why the USSR should sue for peace, especially if the war is going well for them in Central Europe. John Mearsheimer writes:[41]

> the Soviets could accept the SSBN losses and operate on the assumption that shifts in the strategic balance have no political utility. This is a viable strategy as long as the Soviet Union retains a secure assured destruction capability... As long as the Soviets maintain this capability, they can ignore an unfavourable nuclear balance.

Or, as another analyst put it, 'In a wartime scenario where Soviet forces are making significant gains in Western Europe..., it seems unlikely that the loss of some missile submarines would induce the Soviet leadership to halt or withdraw'.[42]

Various critics, in short, openly doubt whether the Maritime Strategy will provide all the benefits the Navy claims for it. There are serious reasons for questioning major elements of the strategy.

(3) Escalation

Opponents argue in addition that the Maritime Strategy, or major elements thereof, may not only be infeasible and ineffective, but also dangerously escalatory. Proponents of the strategy rightly point out the logical tension between these criticisms: if the strategy will fail both operationally and strategically, then how can it be dangerous as well? However, logic, of course, often does not prevail in situations — such as is likely to be the case in a major war between East and West — in which decision-makers will be under great stress, misperceptions and confusion may abound, the fog of war may be thick. It is not at all impossible to end up with a strategy that is both unsuccessful and escalatory. Attempts to implement the strategy, critics fear, have the potential to trigger escalation.

One element of the strategy that raises this concern is the emphasis on quick exploitation of warning, rapid and decisive decision-making, leading to massive forward movement of forces in a crisis. While the Navy believes that such demonstrations of determination will contribute to deterrence in a crisis, those with worries about escalation wonder whether it is desirable or beneficial for crisis management firstly to put decision-makers under pressure for quick decisions and then to send large forces steaming towards and into Soviet maritime bastions. Critics are troubled by several major questions related to this proposed crisis behaviour. Will the Soviets really view such a step as pacifying in the midst of a crisis serious enough to warrant putting US naval forces into a war-fighting position? Will the USSR really stand aside and let US attack submarines pour into their maritime bastion during a peacetime crisis, thereby ensuring that the bastion is in jeopardy and US strength at a maximum if war comes? Or will the Soviets be tempted to prevent the penetration of their bastion, in which case early forward movement of US naval forces will trigger rather than deter war? Would the US want to confront the Soviet leadership with the need in a severe crisis, to make a decision whether or not to allow the penetration of their bastion? In addition, if US attack submarines are deployed in the north Norwegian and Barents Seas, will not US and Soviet submarines then be engaged in a potentially deadly game of cat-and-mouse in a military environment in which there is a tremendous first-strike advantage and in a physical setting (far north and under water) which does not allow close political or operational control? Those who believe that the answers to these questions are (or might be) upsetting rather than reassuring, are likely to conclude, as did John Mearsheimer, that 'Inserting a large number of attack submarines into the Barents Sea during a crisis...would be very dangerous....'[43]

Even more attention has been devoted, however, to the potential escalatory dangers of attacking the Soviet strategic submarine force: in an era in which it grows ever more difficult to ensure the survivability of land-based forces, the Soviets may view their SSBNs as their strategic reserve force. As noted above, the US Navy believes that successful attack on these forces will cause the USSR to seek war termination, while others suggest that it can afford to ignore such attacks. There is, though, a third possible Soviet reaction: it may respond to the significant loss of SSBNs by escalating to the use of nuclear weapons. It might be driven to such a dramatic step if it thought that its land-based forces were highly vulnerable (not such an unreasonable thought in view of the modernisation of US strategic forces, which is bringing into the arsenal the highly accurate MX and Trident II missiles); under these conditions, the destruction of the Soviet SSBN force would mean the demise of the most survivable component of Soviet strategic forces. The Soviets might find that intolerable and be willing to undertake desperate measures to prevent it. Even short of a situation in which Soviet land-based forces are highly vulnerable, it can be argued that if the Soviets care enough about adverse shifts in the nuclear correlation of forces that they might be willing to sue for peace, then perhaps they care enough that they might be willing to escalate.[44] As Barry Posen has written,[45]

> Regardless of whether Western SSNs are in the Barents Sea or nibbling along its edges, whether they are sinking Soviet SSBNs accidentally or deliberately, and whether they are doing so quickly or slowly, the activity is likely to cause considerable disquiet in the Soviet Union. In wartime, such actions are likely to be interpreted in the worst possible light... A Soviet nuclear response cannot be ruled out.

It is, of course, extremely unlikely that the Soviets would initiate a large-scale nuclear attack against the United States in response to the campaign against their SSBNs. What seems less implausible, however, is Soviet tactical use of nuclear weapons in the Northern theatre, against such targets as carrier battle groups, intelligence facilities, key air bases, and so on.[46] The Soviets may believe, or hope, that a limited tactical employment of nuclear weapons in this way can significantly reduce the threat they face in Northern Waters. Such attacks might seem attractive in that use of a small number of nuclear weapons could have a large effect on the military situation, there would be little or no collateral damage, and the Soviets themselves are not highly vulnerable to symmetrical attacks at sea.

Of particular importance with respect to protecting their SSBNs, ASW capabilities appear to degrade dramatically in the face of nuclear attack, not only due to the loss of naval assets such as carriers and attack submarines, but also as a result of the loss of critical but highly vulnerable, command, control, and intelligence capabilities.[47] There is no doubt that the USSR possesses the ability to carry out such attacks, it seems as though doing so might have high utility for the Soviets in terms of protecting greatly valued assets, and there are at least some American analysts who believe that the Soviets will not hesitate in launching a tactical nuclear strike at sea if they think it is necessary. As one detailed analysis of nuclear weapons in Soviet naval doctrine concludes,[48]

> It is quite possible that the Soviet navy will employ nuclear weapons against Western attack submarines should these pose a serious challenge to the security of its SSBN sanctuaries. Nuclear weapons are once again regarded by the Soviet navy as the surest means of carring out a vital mission, and the doctrinal evidence suggests that Moscow will not be reluctant to exercise this option if it is deemed to be critical to the survival of a secure reserve force.

Obviously, those who share this conclusion, and even those who merely fear that it might be correct, will have the gravest doubts about the wisdom of attacking Soviet SSBNs. Such critics will prefer to refrain from undertaking this campaign rather than risk pushing the Soviets across the nuclear threshold.[49]

While the anti-SSBN campaign has attracted the most criticism, forward carrier operations have also raised concern about the risk of nuclear escalation, in part because the carriers themselves are potential nuclear delivery platforms (and are viewed as such by the Soviets) and in part because they are inviting targets for nuclear attack. As one US naval officer wrote in 1985,[50]

> all evidence suggests that the Soviet navy expects to use nuclear weapons, will find it overwhelmingly advantageous to do so, and will probably not be deterred from doing so... Thus, any naval warfighting doctrine must be based on the presumption that nuclear weapons will be used, not on the fervent (and unrealistic) hope that the conflict will stay at the conventional level... The US Navy must recognize the fallacy of deploying its carrier battle groups in the area of most potent Soviet threat and, falsely assuming the Soviets will not use their extensive maritime theatre nuclear warfare

capability, expecting the carrier battle groups to survive. In fact, forward deployment of the carrier battle groups could force the Soviets over the nuclear threshold.

Similarly, Captain Linton Brooks has written that

> carriers approaching within strike range of the Soviet Union may be perceived as preparing to launch nuclear strikes. The Soviets, who stress preemption, may elect to beat us to the draw. Thus forward operations of carrier battle groups seems inconsistent with the...criterion of not inducing escalation.[51]

Again, those who find such analyses to be persuasive will undoubtedly view carrier operations in the far north as being dangerously risky.

In each of these cases, with respect to early forward movement of forces in crisis, to the campaign against Soviet SSBNs, and to forward carrier operations, the large uncertainty that looms over the debate about the dangers of such actions has to do with the most likely Soviet response. How will the Soviets react? That is the key question on which proponents and opponents of the Maritime Strategy divide. It is imperative to note, in considering this debate, that there are few grounds for high confidence in any answer to this pivotal question. The Soviet system is notoriously opaque when it comes to military affairs, and there appears to be little understanding in the West of what thresholds, if any, the Soviets recognise, of how the Soviets view the issue of escalation, and of whether the Maritime Strategy violates, or fails to violate, Soviet sensibilities on this score.[52] Ultimately, whether one views the Strategy as excessively risky or not depends, in simplest terms, on whether one thinks the Soviets are more likely to escalate or to capitulate. While the Navy is confident that, in the event, the Soviets will back down, others are unwilling to risk such enormous stakes on such an uncertain bet.

(4) Alternatives

It is a moot point to raise criticisms, however severe, of a strategy if there are no meaningful alternatives to it. The US Navy argues that the forward Maritime Strategy is the best, and perhaps the only effective, way to perform its essential missions.[53] Critics, naturally, believe that there are acceptable, even preferable, alternatives, and offer two approaches that, while not inevitably linked, are quite complementary.

At the level of overall US military policy, opponents of the Strategy urge the adoption of a strategy that reduces the priority given to the Navy and, in particular, that reduces the Navy's claim on defence resources. The Navy's prosperity in the 1980s, in this argument, has come at the expense of other, more pressing needs, notably the need to enhance NATO's conventional capability in Central Europe. (This clash between those who are preoccupied with Central Europe and those who are committed to the Maritime Strategy has caused this dispute to be characterised as the Continental–Maritime Debate.)[54] Critics suggest that the extremely ambitious and offensive nature of the Strategy has led to a substantial and needless overinvestment in naval power. Brookings Institution defence analyst Joshua Epstein, for example, has written that 'the Maritime Strategy represents a massive diversion of resources from the immediate defense of Western Europe and into offensive surface and subsurface naval actions that are grossly inefficient in accomplishing essential naval tasks..'[55]

The logic of this criticism is twofold. Firstly, as noted above in the discussion of the effectiveness of the Maritime Strategy, many simply doubt that the Soviet Union, a great continental power, is very vulnerable to the exercise of naval power.[56] Furthermore, the Red Army is seen as the primary threat, and it can be effectively negated only by land power. As Robert Komer puts it,[57]

> the kind of carrier-heavy navy we are building and the peripheral maritime strategy for which it is designed, cannot meet our basic strategic needs. Even if we simultaneously swept the Soviets from all the seven seas at the outbreak of a war, this could not alone prevent the USSR from dominating the entire Eurasian land mass... Only land and air power as well could do that.

According to this perspective, naval power should be a secondary priority in a struggle against a major continental power.

The other logical argument in support of the proposition that the Navy's priority in US security policy should be reduced states that not only is the Central European contingency more important, but also that NATO's forces there are more in need of enhancement than is the US Navy. If one believes, as many do, that NATO's forces in Central Europe are inadequate, perhaps woefully so, then it is easy to reach the conclusion that this deficiency should receive higher priority than the Navy: after all, it is commonly remarked, if NATO loses a short war in Central Europe, its naval superiority will be a small solace.

Concerns of this sort have resulted in serious criticism of the naval build-up that has accompanied the Reagan Administration's adoption of the Maritime Strategy. Edward Luttwak writes[58]

> what is extraordinary about the Reagan upsurge is that it has *not* been accompanied by any significant numerical increase in the forces, except for the Navy whose forces are the least useful in countering the strongest part of Soviet military power.

John Mearsheimer also makes this point very emphatically. His conclusion conveys this point so well that it is worth quoting at length:[59]

> The Reagan Administration has not sought to increase the size and strength of NATO's ground and air forces. Instead it has essentially maintained the status quo on the Central Front, even though the Administration is clearly identified with the position that NATO is badly outnumbered in Europe. Perhaps this decision was made on the assumption that substantially increasing the size and strength of the Navy would markedly enhance the allies' deterrent posture in Western Europe. Unfortunately, this assumption is invalid. The best way to achieve that end is to invest more heavily in those forces that stand in the way of the Warsaw Pact armies - NATO's ground and air forces. The Administration passed up this opportunity to devote extra resources to the Navy. Thus, the Administration's defense buildup has done remarkably little to improve NATO's prospects of deterring the Soviet Union in a crisis.

In short, one alternative offered by critics is to spend more on forces for the Central Front and less on the Navy, which would almost surely render infeasible an extremely ambitious doctrine like the Maritime Strategy.

What, then, would become of the US Navy? Here we encounter the second major alternative offered by opponents of the Strategy: with very few exceptions, they argue that a more defensive (and less ambitious) naval strategy would be sufficient to fulfil essential NATO and US naval missions, including notably the protection of sea lanes and the defence of allies. They question whether it is really necessary to chase Soviet submarines in the Barents Sea or to launch carrier strikes against the Kola Peninsula in order to perform these missions. A more defensive strategy, the critics suggest, would be less costly, less escalatory, more realistic in its assumptions about what naval power can contribute in a war against the Soviet Union, and adequate for the defence of NATO's vital maritime interests.

In sum, the Navy's clear and articulate case for the Maritime Strategy has been disputed in several fundamental ways: critics claim that the Strategy is likely to be infeasible, ineffective, and escalatory, and they recommend that alternative strategies be adopted. Two of the key elements of this argument — the issues of effectiveness and of escalatory risks — have to do with predictions of Soviet reactions to the implementation of the Strategy if war comes. Consequently, the controversy has a theological flavour to it, as beliefs about likely Soviet behaviour in this circumstance lead individuals to support or to oppose the strategy. The US Navy expects the Soviet Union to be strongly influenced by its forward operations; critics believe the USSR will be preoccupied with the enormous battle in Central Europe. The Navy is betting heavily on Soviet restraint, believing that the USSR will prefer to back down rather than escalate; its critics fear a more violent Soviet response and dread the disaster such a response might bring. The debate continues, but the policy remains in place.

THE MARITIME STRATEGY AND NORTHERN EUROPE

The adoption of the Maritime Strategy by the United States is clearly a development of great significance for northern Europe, for it ensures that this region will be prominent in the military confrontation between the superpowers. The Strategy, however, is not the cause but rather a consequence of the growing strategic importance of northern Europe. Gradually over a period spanning nearly two decades, this region has been transformed from 'a quiet corner of Europe' into a primary arena for superpower military engagement. Indeed, from an American perspective there is virtually no region in the world that has witnessed more dramatic changes in the politico-military environment than has northern Europe. The nature of Soviet military policy, the evolution of military technology, the American reaction to increased Soviet capabilities in the north, and the inescapable geographical realities of the region have combined to make northern Europe a vital geostrategic component of the clash between the superpowers. The Maritime Strategy guarantees that this will be even more true in the future than in the past.

The process of transformation got underway in the mid-1960s as the Soviets embarked on a major (and, it has turned out, sustained) naval modernisation campaign.[60] As early as 1970 the US Embassy in Oslo was reporting to Washington that the expansion of Soviet naval capability in the north was causing serious concern in Norway.[61] The immediate source of worry for Norway, of course, was the improvement in, and the expanding pattern of activity of, the Soviet Northern Fleet, based on the Kola Peninsula. Viewed in combination with the shrinking US naval presence in the North Atlantic in the 1970s (caused by the conjunction of bloc obsolescence of Second World War-vintage vessels and declining post-Vietnam defence budgets), this development raised the possibility of a significant adverse shift in the naval balance in the North Atlantic. Thus, as Soviet naval capability in the north grew, Western concerns were quickly aroused.[62]

While the improvement in the Soviet Navy did not pass unnoticed in the United States, what really drew the attention of the US Navy northward was the emergence and growth of the Soviet strategic submarine force.[63] Equally important for the structure of security in northern Europe, the Soviets rapidly modernised their sea-launched ballistic missiles (SLBM), increasing their range sufficiently that the Norwegian and Barents Seas were transformed from *transit* areas for Soviet SSBNs heading to firing positions in the Atlantic off the coast of the United States into *deployment* areas for Soviet SSBNs, in which they could be protected and from which they could fire against targets in the United States.[64] Hence, if the US Navy wanted to be in a position at least to threaten, if not destroy, Soviet SSBNs, it would not only have to go north, but it would probably have to fight its way into these waters. (This, of course, is exactly what the Maritime Strategy calls for.)

Thus, for roughly two decades, from the formation of NATO in 1949 until about 1970, unquestioned NATO naval predominance provided confidence that Western interests on the Northern Flank could be defended and there were no compelling reasons for this region to be at the centre of American or NATO defence concerns.[65] Since about 1970 there have been several reasons for steadily growing interest in, and worry about, security in northern Europe: the Soviet acquisition of capability to, at a minimum, raise questions about NATO's ability to protect the North Atlantic sea lanes and to reinforce Norway provided one source of concern. Even more importantly, at least from an American perspective, the deployment pattern of the Soviet SSBN force has caused the High North to become a vital component of *the strategic nuclear equation*. Nothing would be more likely to attract American attention and American military capability than that. Soviet sea-based nuclear forces are the magnet that is drawing the US Navy ever northward.[66]

As a result of these developments, it is clear that the security environment in northern Europe has been irrevocably changed. It is time to set aside the comforting and persistent myth that Scandinavia is a region of stability and low tension. In fact, it is perhaps second only to the Central Front as a region of superpower military confrontation, it is the site of dramatic accumulations of conventional and nuclear military power, and (more than the Central Front) it has a direct connection with the strategic nuclear balance. In any major war, furthermore, this region would be a significant battleground and could easily be the region in which the nuclear threshold is first breached. When one views the changes in the past decade in the behaviour of a number of the major players in the north — the United States, the Soviet Union, Norway and Sweden — stability hardly seems the appropriate word to describe conditions in this region.[67]

It is also very important to realise that the transformation of the High North is not completed: the trend will continue for some time into the future. There are several reasons why this region will continue to grow in importance. Firstly, Soviet land-based strategic forces are becoming, if they are not already, vulnerable. The modernisation of US strategic forces, namely the deployment of the MX missile and the Trident II SLBM, is providing the United States with a growing countersilo capability. The Soviets are, of course, pursuing a mobile land-based missile option, and perhaps this will provide an answer, or a partial answer, to their growing vulnerability problem. However, while the problems of land mobility for ICBMs are not insoluble, they are far from straightforward or easy.[68] Consequently, Soviet land-based forces could, at a minimum, face a period of vulnerability, and that period could turn out to be quite protracted if the Soviets encounter difficulties in achieving a satisfactory land mobile system. To the extent that this is true, the importance of Soviet sea-based strategic forces will increase accordingly. In the future, the Soviet SSBN force may be the most survivable, if not the only survivable, component of their strategic forces.[69] Should this come to pass, it will surely cause the Soviet interest in their Northern Waters bastion to become even more acute.

Secondly, the USSR is in the process of placing multiple warheads (MIRVs) on its SLBMs. With the arrival of the new generation of MIRVed SLBMs (the SS-N-20), this is changing rapidly. As the Soviets put MIRVed missiles out to sea, the sea-based force will account for a growing proportion of Soviet nuclear capability.

A more troubling possibility, though, is that in the aftermath of President Reagan's renunciation of SALT, the Soviets will no longer consider themselves bound by the agreements, in which case they can deploy as many MIRVed SLBMs as they care to produce.[70] Obviously, the more the USSR proliferates SSBNs, the greater will be both Soviet and American interest in the Soviet SSBN bastions.

Thirdly, it seems probable that the Soviet Union will increasingly exploit the Arctic Ocean as a deployment area for its strategic submarines. The advantages of so doing were described in a recent article on under-ice operations:[71]

> Deploying in the Arctic, Soviet submarines have no chokepoints to transit, and their operating areas are considerably more familiar to them than to the West; conversely, Western anti-SSBN submarines have themselves to transit the chokepoints of the Davis and Denmark Straits; the areas are great distances from the US and UK bases, and the inventory of knowledge of the operating areas is relatively poor... Further, a major contribution to Western ASW capabilities has been their highly superior air and surface assets, and the Arctic icecap effectively rules them out as nonplayers in the new arena.

In addition, the Arctic is a very noisy environment, due in part to the noise generated by the ice, and this hampers US ASW sensors: according to one report, the noise level is such that it 'all but drowns out the faint sound of a submarine'.[72] Apparently, the USSR has not to date made major use of the Arctic as an SSBN deployment area, but it is evident that it has every incentive to do so.[73] It also appears that it will have increasing capability to do so: 'Enormous Soviet effort has gone into developing longer range SLBMs and submarines that are specifically designed to operate for extended periods under the ice.'[74] One example is the new SSBN, the Typhoon, which is thought to have been designed with under-ice operations in mind. As the Soviet Navy moves under the ice, there can be no doubt that the US Navy will be pulled into the Arctic as well. Already there are clear signs that this is happening. Beginning in the early 1980s, for example, the US Navy began conducting Arctic anti-submarine warfare exercises;[75] and it has already begun funding the development of a new attack submarine (the SSN 21) which is intended to give the United States the ability to threaten new generation Soviet submarines. Thus, the superpower game in the High North is going to be pulled even further north, and the Arctic Ocean and adjacent littoral territories are going to become prime geostrategic regions.[76]

For NATO's northern members, Norway, Greenland and Canada, this is a development of utmost importance, for it implies not only expanded Soviet and American military activities in the High North, but also new need for intelligence, command and control, and logistical facilities.[77] This is likely to raise new issues on the defence policy agendas of these countries — indeed, it has already begun to do so. It also means that legal, diplomatic and economic issues relating to the Arctic will be even more sensitive and more fraught with implications for security policy than has been the case in the past.

Finally, but not least among these concerns, the emergence of the Arctic as an important strategic area will raise questions about whether NATO as presently structured is configured to address Arctic issues adequately. NATO's four 'Arctic' members (including the United States, by virtue of the location of Alaska) do not presently have a mechanism for jointly considering the panoply of problems and controversies (including possible conflicts of interest among them on matters such as resource exploitation and environmental protection) that confront them in this region. Diplomatically, is the network of bilateral relationships and arrangements between the United States and Norway, Greenland (Denmark), and Canada enough or should there be more collaboration and co-ordination among these four states? Militarily, is the present NATO command structure appropriate for the effective handling of Arctic security issues? Does the jurisdiction of the Supreme Allied Commander Atlantic (SACLANT), for example, extend to the entire Arctic, in which case he may have submarines operating in the Bering Strait area in the Far East? What happens if forces from the US Pacific Fleet venture into the Arctic? Who commands them? Is there adequate communication between the Pacific Fleet and SACLANT so that American submarines do not end up stalking other American submarines? If there is interest in pursuing Polar arms control, how is NATO going to organise itself for such negotiations? Is the Headquarters, Allied Forces Northern Europe (AFNORTH), devoted as it is to the defence of Norway and Denmark, capable of incorporating into its mandate Canada's considerable Arctic concerns? Or, to cope with matters such as these, does NATO need to consider setting up an Arctic command? However unclear the answers may now seem, the time has certainly come to be asking such questions. The Arctic is becoming, as the Norwegian and Barents Seas have already become, an important arena for superpower military competition.

In combination with these other factors — the growing size and importance of the Soviet SLBM force and the increasing strategic significance of the Arctic — the Maritime Strategy makes it certain that northern Europe will be a major point of confrontation between the superpowers. It proclaims that the US Navy *will* go north in pursuit of the Soviet Navy, that it *will* attempt to shatter the Soviet bastion, that it *will* seek out and destroy Soviet SSBNs wherever they may be. This means, of course, that the High North will be the scene of substantial military deployments in peacetime and of major naval (and, most likely, land) battles in the event of war. The only thing that could really prevent this would be an agreement between the superpowers to turn the Northern Waters into an SSBN sanctuary into which US forces would not venture, but there is no sign that this is even remotely in prospect, nor is it likely ever to be. A more defensive US naval policy, particularly one that eschewed the campaign against Soviet SSBNs, could significantly reduce, although it would not eliminate, the impact of these trends on northern Europe. However, this is not a likely possibility either: not only is the US Navy firmly committed to offensive doctrines, but it would require the overturning of four decades of US strategic nuclear doctrine for it to refrain from seeking to destroy Soviet SSBNs. In any case, under the current Administration, US military policy has moved in the other direction, towards *greater* commitment to offensive naval strategy and to strategic counterforce. The steps that might insulate northern Europe from the implications of its growing strategic importance are very unlikely indeed.

One is left, then, with the conclusion that northern Europe, far from being a region of stability, is a very dynamic strategic and political–military environment. The Maritime Strategy is prominent among the factors contributing to this dynamism, but it is neither the only nor the primary cause; developments in Soviet defence policy are the root of the problem. When the strategic situation does stabilise, it is likely to be stability like that which exists in central Europe: tense, heavily armed, and a chronic source of concern. No longer does this region enjoy the blessing of being a quiet corner of Europe.

NOTES

1. The US Navy came to believe that misunderstanding of the Maritime Strategy resulted in inaccurate and unfair criticisms of it. See, for example, Major Hugh K. O'Donnell, Jr, 'Northern Flank maritime offensive', *US Naval Institute Proceedings*, September 1985, p. 49. For discussion of the ambiguous public character of the Maritime Strategy, see Stanley J. Heginbotham, 'The Forward Maritime Strategy and Nordic Europe', *Naval War College Review*, November–December 1985, pp. 22–4.

2. See for example, Admiral Thomas B. Hayward, 'The future of US sea

power', *US Naval Institute Proceedings*, May 1979, pp. 66–71. On Navy friction with the Carter Administration, see for example, O. Kelly, 'US Navy in distress', *US News and World Report*, 6 March 1978, pp. 24–6; and 'Navy protests limitation of its long-term mission', *New York Times*, 14 March 1978. The Navy objected to the Carter Administration's opposition to new carriers and to its emphasis on smaller, less capable vessels. A recent article dates the origins of the Maritime Strategy concept to the mid-1970s. See William L. Chaze, 'Rust to riches: the Navy is back', *US News and World Report*, 4 August 1986, p. 29.

3. John F. Lehman, Jr, 'Rebirth of a US naval strategy', *Strategic Review*, Summer 1981, p. 13. For another early indication of Lehman's views, see 'Lehman seeks superiority', *International Defense Review*, May 1982, pp. 547–8.

4. 'Unclear sailing', *Newsweek*, 16 March 1981, p. 24 (emphasis added). For more on the early strategic inclinations of the Reagan Administration and their implications for the Norwegian Sea, see Steven E. Miller, 'The northern seas in Soviet and US strategy' in S. Lodgaard and M. Thee (eds), *Nuclear disengagement in Europe*, (Taylor and Francis, London, 1983), pp. 127–30.

5. For a very brief but informative account of the evolution of the Maritime Strategy, see Captain Peter M. Swartz, 'Contemporary US naval strategy: a bibliography' in *The Maritime Strategy*, supplement to *US Naval Institute Proceedings*, January 1986, pp. 41–2; House Armed Services Committee, *Hearings on the Department of Defense Authorization for FY 1984* (part 4), (USGPO, Washington, DC, 1983), pp. 47–51.

6. Senate Armed Services Committee, Subcommittee on Sea Power and Force Projection, *Department of Defense Authorization for Appropriations for Fiscal Year 1985: Maritime Strategy* (USGPO, Washington, DC, 1985).

7. See, for example, David Rivkin, 'No bastions for the bear', *US Naval Institute Proceedings*, April 1984, pp. 36–43; versus Richard Ackley, 'No bastions for the bear: round 2', *US Naval Institute Proceedings*, April 1985, pp. 42–7. Also see, for example Captain Linton Brooks, 'Escalation and naval strategy', *US Naval Institute Proceedings*, August 1984, pp. 33–7.

8. These questions will be dealt with in a subsequent section.

9. A further thoughtful articulation and defence of the Maritime Strategy is Captain Linton Brooks, 'Naval power and national security: the case for the Maritime Strategy', *International Security*, Fall 1986, pp. 58–88.

10. Ibid., pp. 59–60.

11. These exercises have been extensively publicised. See, for example, Rodney Cowton, 'NATO Fleets go on offensive', *The Times* (London), 30 August 1985; 'US carriers in Arctic show of force', *The Times*, 29 August 1985; 'US–Soviet collision course in Seas North of Norway', *Washington Times*, 16 September 1985. Jack Dorsey 'NATO forces flex muscles in Norwegian Sea', *Norfolk Virginian-Pilot*, 9 September 1985.

12. Jack Dorsey, 'NATO pushes exercises into Norwegian Sea', *Norfolk Ledger-Star*, 9 September 1985.

13. See, for example, Paul Bedard, 'Naval air: strategy changes from a defensive role to projecting power into Soviet backyard', *Defense Week*, 1 July 1985, pp. 15–16.

14. Admiral James D. Watkins, 'The Maritime Strategy' in *The Maritime Strategy* (supplement), pp. 4, 16.

15. Ibid., p. 7 (emphasis in original).
16. John Lehman, 'The 600-ship navy' in *The Maritime Strategy* (supplement), p. 37.
17. Brooks, 'Naval power', p. 66.
18. This brief description is drawn from the following sources: Watkins, 'The Maritime Strategy'; O'Donnell, 'Northern Flank'; and John J. Mearsheimer, 'A strategic misstep: the Maritime Strategy and deterrence in Europe', *International Security*, Fall 1986, pp. 3–57. Those wishing a more detailed account of the strategy should consult these sources.
19. It is often remarked that by operating offensively far forward, the US Navy puts the Soviet Union Northern Fleet on the defensive and protects the SLOC by forcing the Soviet Northern Fleet to 'stay at home' in order to protect its SSBNs. Little attention has been paid to the reverse logic: the USSR may be able to reduce the pressure against its northern seas bastion by vigorously attacking the sea lanes. This could force the US to hold some of its forces back for the SLOC battle rather than send them forward against the Soviet bastion. It is possible that the USSR could pre-deploy attack submarines in the North Atlantic during a crisis, thereby augmenting its ability to pose a serious threat by making the transit through NATO anti-submarine warfare (ASW) capability in the Greenland–Iceland–United Kingdom (GIUK) Gap *before* the war has started. This scenario raises the prospect that, in a crisis, US attack submarines would be surging northward while Soviet attack submarines would be surging southwards. The key question in such a situation is: whose forces will be most tied down by the operations of the other? *If* Soviet SSBNs represent redundant strategic nuclear capability, while NATO's SLOCs are vital, the answer to this question is unlikely to be favourable to the US and NATO.
20. It should be recognised, however, that the refuelled range of carrier-based attack aircraft is quite extensive and carriers do not need to operate very far forward in order to be within range of targets on the Kola Peninsula.
21. Watkins, 'The Maritime Strategy', p. 9. See also Brooks, 'Naval power', p. 65.
22. Watkins, 'The Maritime Strategy', p. 12.
23. On these points, see, for example, the article by Vice Admiral Henry Mustin (Commander, Atlantic Strike Fleet), 'The role of the Navy and Marines in the Norwegian Sea', *Naval War College Review*, March–April 1986. The ability of carriers to utilise speed, weather, electronic war and emission control to 'hide' themselves should not be overlooked or underestimated. See, for example, 'How to make carriers vanish', *Journal of Commerce*, 20 August 1986, p. 20. See also Mark Cotton, 'Naval electronic warfare: enhancing warship survivability', *International Defense Review*, Special Supplement no. 2, (1985), pp. 24–7.
24. On Mahan's views and influence, see Margaret Sprout, 'Mahan: evangelist of seapower' in Edward Meade Earle (ed.), *Makers of modern strategy* (Princeton University Press, Princeton, 1943). For a very useful overview of the evolution of the US Navy, see Kenneth J. Hagan (ed.), *In peace and war: interpretations of American naval history, 1775–1978* (Greenwood Press, Westport, Conn., 1978). For an overview of US naval policy since the Second World War, see Steven E. Miller, 'The US Navy in the nuclear age' in Michael Mandelbaum (ed.), *Postwar US military policy*, forthcoming.

25. See, for example, Barry R. Posen, 'Inadvertent nuclear war? Escalation and NATO's Northern Flank' in Steven E. Miller (ed.), *Strategy and Nuclear Deterrence*, (Princeton University Press, Princeton, 1984), pp. 96–106, for clear indication that the commitment to forward operations antedates the Reagan Administration.

26. Ibid., pp. 97–8.

27. See, in particular, David Alan Rosenberg, 'The origins of overkill: nuclear weapons and American strategy, 1945–1960', and Desmond Ball, 'US strategic forces: how would they be used?', both in Miller (ed.), *Strategy and nuclear deterrence*, pp. 113–82, 215–44.

28. F.J. West, Jr, 'Maritime strategy and NATO deterrence', *Naval War College Review*, September/October 1985, p. 12.

29. For an extensive and thorough exposition of most of the major criticisms, see Mearsheimer, 'A strategic misstep'. O'Donnell provides a useful summary and rebuttal of the main criticisms in 'Northern Flank maritime offensive'. Brooks provides a more elaborate rebuttal in 'Naval power'. Further defences of the Maritime Strategy are Colin S. Gray, 'Maritime Strategy', *US Naval Institute Proceedings*, February 1986, pp. 34–42; and John Allen Williams, 'Rethinking the Forward Maritime Strategy' (Loyola University of Chicago, December 1985).

30. For a concise (but thorough) sceptical view, see Tom Stefanick, 'Attacking the Soviet sea-based deterrent: clever feint or foolhardy maneuver?', *F.A.S. Public Interest Report*, June–July 1986. See also Stefanick's book-length study of ASW, *Strategic antisubmarine warfare and naval strategy* (Lexington Books, 1987). Also very informative on these issues is the work of Donald Daniels. See his 'Antisubmarine warfare in the nuclear age', *Orbis*, Fall 1984, pp. 527–52; and *ASW and superpower stability*, forthcoming from IISS. For a recent brief survey of ASW developments, which suggests that 'the hunters may actually be losing ground', see Malcolm W. Browne, 'In battle of wits, submarines evade advanced efforts at detection', *New York Times*, 1 April 1986. For doubts about US under-ice ASW capability, including at least one opinion that Soviet subs under the ice are 'invulnerable', see Barry Brown, 'Growing Soviet sub force in Arctic Ocean worries US', *Washington Times*, 7 August 1985.

31. Mearsheimer, 'A strategic misstep', p. 50. I have heard this view expressed in private by naval officers, which leads me to believe that a minority *within* US and Allied navies share this conclusion.

32. Jeffrey Record, 'Jousting with unreality: Reagan's military strategy' in Steven E. Miller (ed.), *Conventional forces and American defense policy*, (Princeton University Press, Princeton, NJ, 1986), pp. 70–3.

33. Stansfield Turner and George Thibault, 'Preparing for the unexpected: the need for a new military strategy', *Foreign Affairs*, Fall 1982, p. 127.

34. William F. Kaufmann, *The 1986 Defense Budget*, (The Brookings Institution, Washington, DC, 1985), p. 35. For a more recent severe attack, see Joshua M. Epstein, *The 1987 Defense Budget*, (The Brookings Institution, Washington, DC, 1986), pp. 41–2.

35. Watkins, 'The Maritime Strategy', p. 4.

36. Senate Armed Services Committee, *Department of Defense Authorization*, p. 3893.

37. Mustin, 'The role of the Navy', p. 3.

38. See, in particular, Mearsheimer, 'A strategic misstep'.

39. Robert W. Komer, 'Maritime strategy vs. coalition defense', *Foreign Affairs*, Summer 1982, p. 1133. Komer elaborated on his views in his book, *Maritime strategy or coalition defense?* (Abt, Cambridge, MA, 1984), pp. 39–76. On this point see in particular p. 67.

40. Edward N. Luttwak, *The Pentagon and the art of war: the question of military reform* (Simon and Schuster, New York, 1985), p. 64.

41. Mearsheimer, 'A strategic misstep', p. 52.

42. Stefanick, 'Attacking the Soviet sea-based deterrent', p. 5.

43. Mearsheimer, 'A strategic misstep', p. 46. See also Posen, 'Inadvertent nuclear war?', and Stefanick, 'Attacking the Soviet sea-based deterrent'.

44. For fuller discussion of these issues, see the sources cited in note 43.

45. Posen, 'Inadvertent nuclear war?', pp. 43–4.

46. For discussion of these issues, see Desmond Ball, 'Nuclear war at sea', *International Security*, Winter 1985/1986, pp. 8–31. See also Charles Stafford, 'A nuclear war might well be spawned at sea', *St Petersburg Times*, 2 June 1986. Also relevant is John Ausland, 'The silence on naval nuclear arms should be broken', *International Herald Tribune*, 12 March 1986. For an important early discussion of this question, see Linton Brooks, 'Tactical nuclear weapons: the forgotten facet of naval warfare', *US Naval Institute Proceedings*, January 1980, pp. 28–33. Also useful is Gordon McCormick and Mark Miller, 'American seapower at risk: nuclear weapons in Soviet naval planning', *Orbis*, Summer 1981, pp. 351–69.

47. See, for example, Posen, 'Inadvertent nuclear war?', p. 43. Also Richard Garwin, 'The interaction of anti-submarine warfare with the submarine-based deterrent' in K. Tsipis, A. Kahn and B. Feld (eds), *The future of the sea-based deterrent*, (MIT Press, Cambridge, 1973), p. 89.

48. McCormick and Miller, 'American seapower at risk', p. 359.

49. It is important to emphasise that the issue here is not simply to avoid being provocative to the USSR: all this will be taking place, after all, in the midst of a major, and surely very violent, war. Rather, the point is to ensure that US (and NATO) behaviour does not trigger a form of this war — in particular, a nuclear war of any variety — that the US and NATO do not want to fight.

50. Commander Raymond E. Thomas, 'Maritime theater nuclear warfare: matching strategy and capability' in *Essays on strategy* (National Defense University Press, Washington, DC, 1985), p. 50.

51. Linton F. Brooks, 'Escalation and naval strategy', *US Naval Institute Proceedings*, August 1984, p. 36. Brooks also notes the potential role of carriers as nuclear attack platforms in his 'Naval power', p. 73.

52. For a rare analysis of Soviet views of escalation, see Joseph Douglass, Jr and Amoretta Hoeber, *Conventional war and escalation: the Soviet view* (Crane Russak, New York, 1981), which argues that the USSR has *not* shifted to the expectation of protracted conventional war, but rather believes the nuclear phase of any large war would be decisive. Obviously, this implies that the Soviets will not go to great lengths to keep the war conventional, which could lead to the conclusion either that the Maritime Strategy is very dangerous because the Soviets intend to escalate or not dangerous at all because the Soviets intend to escalate anyway. The analysis, however, does not directly address naval issues. See also R. Welander, J. Herzog and F. Kennedy, *The*

Soviet Navy declaratory doctrine for theater nuclear warfare (BDM Corporation (for the Defense Nuclear Agency), Washington, DC, 1977), which attempts to assess Soviet views of escalation in a naval context but complains (p. 15) of 'the apparent inability of Soviet military theoreticians to come to grips intellectually with the concept of escalation'. The analysis finds no clear indication of whether and when the USSR might use nuclear weapons at sea, but notes that Soviet military doctrine in general is shaped by the expectation that the war will be nuclear. Michael MccGwire shares this conclusion: see his 'Soviet military thought: its implications for maritime warfare', unpublished paper, (Brookings Institution, June 1983), p. 11.

53. See, for example, Robert S. Wood and John T. Hanley, Jr, 'The maritime role in the North Atlantic', *Naval War College Review*, November–December 1985, pp. 5–18, which specifically rejects the successful defensive sea control strategy of the Second World War as inadequate and inappropriate to current requirements of the US Navy.

54. For a full discussion of the issues raised in this debate, see Keith A. Dunn and William O. Staudenmaier, *Strategic implications of the continental–maritime debate*, (Praeger, New York, 1984; The Washington Papers/107).

55. Epstein, *The 1987 Defense Budget*, p. 44. On the same point, see also Luttwak, *The Pentagon*, p. 64.

56. This point is sometimes seen in terms of the confrontation between Mahan's theory of naval primacy and Mackinder's geopolitical arguments about the predominant position of the Eurasian heartland. Many now believe that Mahan was wrong and Mackinder correct. See, for example, Mearsheimer's discussion of 'Sea power in the industrial age' in 'A strategic misstep', pp. 32–5. For a superb discussion of Mahan versus Mackinder, see Paul M. Kennedy, *The rise and fall of British naval mastery*, (Macmillan Press, London, 1983), pp. 177–202.

57. Komer, *Maritime strategy'*, p. 74.

58. Luttwak, *The Pentagon*, p. 256.

59. Mearsheimer, 'A strategic misstep', pp. 55–6.

60. For a brief overview, see Steven E. Miller, 'Assessing the Soviet Navy', *Naval War College Review*, September–October 1979.

61. See John Ausland, 'Norwegian concern with growing Soviet naval threat to Norway' (Department of State Telegram Oslo 04416 1117072 (Declassified), 11 December 1970). I am grateful to John Ausland for bringing this document to my attention. An article in *NATO Letter*, September 1970, which documented the expansion of Soviet naval activities in the Norwegian Sea, was the catalyst that crystalised Norwegian concerns about the adequacy of their defence arrangements in the face of the growing Soviet threat.

62. See J.J. Holst, 'The Soviet buildup in the Northeast Atlantic', *Survival*, January–February 1972, p. 25. See also the very useful survey of developments by General Tonne Huitfeldt, 'The Maritime environment in the North Atlantic' in *Power at sea, III: competition and conflict*, Adelphi Paper no. 124 (1976), pp. 13–21. An indication of the growing interest in, and concern about, the Northern Flank is Christoph Bertram and Johan Holst, *New strategic factors in the North Atlantic* (IPC Sciences and Technology Press, Guildford, UK, 1977). Throughout the mid-1970s, the changing military situation in the north attracted considerable notice in the press. See, for example, Ludovic Kennedy,

'The Soviet Fleet casts a long shadow over Norway's border', *The Times*, 5 November 1974; 'Oslo concerned at Soviet buildup', *International Herald Tribune*, 21 May 1975; Desmond Wettern, 'North flank of NATO the weak link', *Daily Telegraph* (London), 26 February 1976; Michael Getler, 'NATO's north flank worried by Soviet, East Bloc buildups', *International Herald Tribune*, 8 May 1976.

63. I discuss this at greater length in Miller, 'The northern seas in Soviet and US strategy' in S. Lodgaard and M. Thee (eds), *Nuclear disengagement in Europe* (Taylor and Francis, London, 1983), pp. 121–2.

64. See, in this regard, Michael MccGwire's important article, 'The rationale for the development of Soviet seapower', *US Naval Institute Proceedings*, May 1980, pp. 155–83. For a concise analysis of Soviet submarine development, see Ian Bellany, 'Sea power and Soviet submarine forces', *Survival*, January/February 1982, pp. 2–8.

65. Here I leave aside the question of the role of Northern Waters as a staging area for early-generation American SSBNs and for American carriers in their nuclear attack role.

66. I have provided here only a brief treatment of these points. For a fuller discussion, see the excellent analysis by Sverre Jervell, *Soviet military buildup in the High North and the Western response: challenges for Norway* (Center for International Affairs, Harvard University, June 1985). Also useful is Nils Petter Gleditsch, 'The strategic significance of the Nordic countries', *PRIO Report* (Peace Research Institute, Oslo), no. 14-85 (December 1985). A very interesting collection of essays that examines economic as well as military factors in the growing geostrategic importance of the High North is Clive Archer and David Scrivener (eds), *Northern Waters: security and resource issues*, (Croom Helm, London, 1986). See, in particular, the chapter by Geoffrey Till, 'Strategy in the far north', pp. 69–84. For an earlier broad overview, not limited to naval matters, see Nils Andren, 'Changing strategic perspectives in northern Europe' in Bengt Sundelius (ed.), *Foreign Policies of northern Europe* (Westview Press, Boulder, Colorado, 1982), pp. 73–106. For an analysis of the increasing military challenge for the US in the north, see Kenneth A. Myers, 'US power projection in the Northern Flank', in Uri Ra'anan, Robert L. Pfaltzgraff, Jr and Geoffrey Kemp (eds), *Projection of power: perspectives, perceptions, and problems* (Archon Books, Hamden, Conn., 1982), pp. 187–206. See Nils Petter Gleditsch, 'Invictus agreement declassified', *PRIO Inform* (Peace Research Institute, Oslo), no. 14-84 (December 1984). For an account of the prepositioning agreement, see Robert K. German, 'Norway and the Bear: Soviet coercive diplomacy and Norwegian security policy', *International Security*, Fall 1982, pp. 72–5.

67. For an excellent brief overview of the submarine incidents and their implications, see Carl Bildt, 'Sweden and the Soviet submarines', *Survival*, July/August 1983, pp. 165–9. See also Kirsten Amundsen, 'Soviet submarines in Scandinavian waters', *The Washington Quarterly*, Summer 1985, pp. 111–22; and General Stig Lofgren, 'Soviet submarines against Sweden', *Strategic Review*, Winter 1984, pp. 36–43. It should be noted that Sweden's defence policies were under pressure even before the submarine incidents. See, for example, Steven Canby, 'Swedish defense', *Survival*, May–June 1981, pp. 116–23. See also Nils Andren, 'Sweden's defence doctrines and changing threat perceptions', *Cooperation and Conflict*, vol. XVII (1982), pp. 29–39.

68. See, for example, the extensive analysis in Congressional Budget Office, *MX Missile basing* (USGPO, Washington, DC, 1981), which concluded that none of the numerous land mobile concepts was attractive for MX. See p. 257.

69. Of course, given the confidence of the US Navy that it can impose severe attrition on the Soviet SSBN force, one may wonder how survivable *it* is. When deployed at sea, however, it is surely *not* vulnerable to prompt destruction, as are land-based forces, and in the future Soviet improvements in quieting their submarines may render them less vulnerable, as may the exploitation of the Arctic Ocean as a deployment area.

70. On Reagan's rejection of the SALT agreements, see, for example, Michael R. Gordon, 'Reagan and Arms Treaty: a sharp shift in policy', *New York Times*, 30 May 1986.

71. See Norman Polmar, 'Sailing under ice', *US Naval Institute Proceedings*, June 1984; Hamlin Caldwell, 'Arctic submarine warfare', *The Submarine Review*, July 1983; T.M. LeMarchand, 'Under ice operations', *Naval War College Review*, May–June 1985, pp. 19–27 — here, p.22.

72. Malcolm W. Browne, 'In battle of wits, submarines evade advanced efforts at detection', *New York Times*, 1 April 1986, quoting Dr Robert Spindel of the Woods Hole Oceanographic Institution.

73. Barry Brown, 'Growing Soviet sub force in Arctic Ocean worries US', *Washington Times*, 7 August 1985, reports from the Pentagon to the effect that the Soviets have not in the past used the Arctic extensively but that they are now operating more submarines in that ocean than ever before.

74. Le Marchand, 'Under ice operations', p. 22. On the need for specially designed equipment in order to operate effectively in the Arctic, see p. 26.

75. Senate Armed Services Committee, *Maritime Strategy*, p. 3883. (From the testimony of Admiral Watkins.) Secretary Lehman has stated that the US Navy will attack Soviet SSBNs 'wherever they may be deployed'. See Melissa Healy, 'Lehman: we'll sink their subs', *Defense Week*, 13 May 1985, p. 13.

76. One of the first to recognise that this would happen was Willy Østreng. See his 'The strategic balance and the Arctic Ocean: Soviet options', *Cooperation and Conflict*, no. 1 (1977), pp. 41–62. For a thorough discussion of the legal and arms control implications, see Lincoln P. Bloomfield, 'The Arctic: the last unmanaged frontier', *Foreign Affairs*, Fall 1981, pp. 87–105.

77. On Canada, see Nicholas Tracy, 'Canada's security considerations in the Arctic' in Archer and Scrivener, *Northern Waters*, pp. 146–54. Evidence of Canadian concern about military capabilities in the Arctic is beginning to appear. See, for example, Standing Senate Committee on Foreign Affairs, Subcommittee on National Defense, *Canada's Maritime Defense* (Ottawa, 1983), especially pp. 50–3. See also Paul Mann, 'Canada's defense policies: budget deficits constrain Canadian defense efforts', *Aviation Week and Space Technology*, 30 June 1986, pp. 44–7, which notes that pressure to extend military commitments to the Arctic are coming at a time when the Canadian government is already feeling overcommitted. On Greenland, see Clive Archer, 'Greenland and the Atlantic Alliance', *Centrepiece no.7*, (Centre for Defence Studies, University of Aberdeen, December 1985). There has been recurrent interest in the possibility that arms control — namely, some sort of Polar Basin agreement — could prevent the militarisation of the Arctic. For a very useful overview, see Ronald G. Purver, *Arms control in the north* (Centre for

International Relations, National Security Series no. 5/81, Queen's University, 1981). Purver's superb footnotes are an excellent short-cut for anyone who wishes to inform themselves on this subject. A related idea is the notion of SSBN sanctuaries, from which hostile forces would be excluded by agreement. See, for example, B. Feld and G. Rathjens, 'Antisubmarine warfare and the sea-based deterrent — opportunities for arms control?', *Survival*, November–December 1973, pp. 268–74; Joel S. Wit, 'Some proposals for controlling antisubmarine warfare' in W. Epstein and B. Feld (eds), *New directions in disarmament*, pp. 77–86; and Ronald G. Purver, 'The desirability and feasibility of negotiated controls on strategic anti-submarine warfare' (International Studies Association, March 1982).

11

Maritime Strategy in Northern Waters: Implications for the Navies of Europe

Geoffrey Till[1]

Previous chapters (6 and 7) have dealt with the Soviet maritime forces in Northern Waters and the strategy adopted by the Soviet Union, as it affects the area. Chapters 9 and 10 have examined the American response to that presence — and indeed to pressures within the United States. It should be remembered, however, that this strategic shadow-boxing between the two superpowers concerns the seas off Europe and naturally European governments, and their navies, have an interest in all these activities. This chapter examines the response of West European states to the emerging maritime confrontation in Northern Waters and will refer especially to their reaction to the American Maritime Strategy.

Current American thinking on maritime strategy has been conveniently summarised for us in a special supplement entitled 'The Maritime Strategy' which appeared with the January 1986 edition of the *Proceedings of the United States Naval Institute*. An accompanying note by the Journal's editor explained that one of the reasons for producing an unclassified version of what the US Navy took to be its part in America's national strategy was the extent to which 'misinterpretations or exaggerations' had affected public debate.[2] While doubtless the editor mainly had the American defence community in mind, the same point has some application in Europe.

European reactions to the US Navy's Maritime Strategy have ranged from warm support to downright hostility. The diversity of response partly reflects differing levels of sympathy for the general strategic aims of the United States, or more particularly of the Administration of President Reagan. In some quarters 'Forward Operations' are seen as all of a piece with President Reagan's operations against Libya, with his enthusiasm for SDI, with his apparently uncompromising attitude to the Soviet Union and with his support of disreputable sets of people in Central America. Others, of course, are much more supportive. The point is that these views are essentially political, but they often have a considerable bearing on the way the Maritime Strategy is perceived and responded to. In short, European reactions to US Navy strategy often tend to reflect differing assumptions about American foreign and defence policy as a whole.

However, the diversity of response is also a function of different ways of interpreting what the Maritime Strategy actually means. Does it mean chasing SSBNs under the Arctic ice and starting off the Third World War, simply maintaining an American naval presence in the Norwegian Sea, horizontal escalation or using aircraft carriers to flatten Murmansk on the first day of the war? Europeans are not helped in this regard by the fact that the policy has changed over the years, and that it sounds rather different when expressed by different types of people.

So this attempt to review the European response to the Maritime Strategy and to assess what its implications for the navies of Europe might be, has to start by identifying what the essential characteristics of the Maritime Strategy are. Broadly, four points need to be made about it.

- The Maritime Strategy is essentially an expression of the maritime component of the national strategy of the United States. It makes no pretentions to being a NATO document.
- The Maritime Strategy is clearly global in perspective, rather than merely European. Its very first sentence reads: 'The United States is inevitably a maritime nation, and the United States and its Navy have inescapable global responsibilities'. Emphasis is given to 'the volatility of today's international situation' and the need to apply the United States' maritime forces of presence and crisis response 'in an expanding set of the world's trouble spots'.[3]
- The Maritime Strategy is presented as a strategy of deterrence. 'Deterrence', it says, 'simply means convincing a potential aggressor that the risks involved in aggression are greater than the possible benefits. The Soviets, or any other potential aggressors, will not be deterred by empty threats and rhetoric. A credible deterrent must have ready and capable forces behind it and the commitment to use them is necessary'.[4]
- Finally, the Maritime Strategy provides a broad and generalised overview of the place of maritime forces in national strategy. Likened by the editor to the British Defence White Paper (in a way which considerably flatters the latter!), it no more goes into precise

descriptions of how the US Navy will conduct itself in certain operational circumstances than does Admiral Sergei Gorshkov's book *The sea power of the state*[5] for the opposition. War-gamers looking in the Maritime Strategy for an explanation of how a carrier battle group would flatten Murmansk will be disappointed.

This review of the implications of the Maritime Strategy for the navies of Europe will therefore concentrate on these four topics in turn.

A NATIONAL STRATEGY

The United States has certain strategic interests in Europe's Northern Waters which derive at least as much from her own basic requirements as they do from those of her allies in Western Europe. A good deal of her interest in this area springs from national perspectives about the strategic nuclear balance between the superpowers, as Tomas Ries's chapter in this book makes clear.

United States' interests in, and support for, anti-SSBN operations is a case in point. The United States has an enormous strategic nuclear arsenal; the present Administration at least has high hopes of being able to deploy an effective strategic defence system; and finally, the US Navy has an ASW capacity which in size and technological sophistication surpasses anything else deployed in the West. All these points give Americans special interests in anti-SSBN operations which Europeans cannot share.

For a start, there are only two European navies capable of going after Soviet SSBNs in any serious way, Britain and France. Yet, however successful they might hope to be tactically, their efforts are very unlikely indeed to make any really significant alteration to the strategic balance between their two countries and the Soviet Union, or to be able to threaten to deprive it of its second strike capability against them (not least because of the relative size of the Soviet strategic arsenal, and its theatre nuclear forces). The rationale for the British and French navies engaging in such operations, therefore, would be rather different from that of the United States. The remaining dozen European navies have no such option anyway.

In the field of conventional operations, too, the United States has distinctively national interests which frame her maritime policy in the north and which would apply whatever the navies of her European allies were doing. No matter what happened to Western Europe, for example, the United States is bound to regard the islands of the North

Atlantic as stepping stones for an eventual Soviet assault on the American continent, or a potential route back should Western Europe be lost, just as she did the islands of the Pacific in the Second World War. Just as it was then, the United States is a long way from the battlefront, and so unlikely to be overwhelmed in the kind of rapid conventional assault for which, we are told, Soviet forces are designed.[6] This is why there is so much more stress on the long war in American than there is in European thinking, and why the United States has continuous interests in the islands of the North Atlantic. Because they are largely the product of geography, these interests are permanent, and moreover, being distinctively American, for the same reason, they will have significant effects on US naval thinking.

Operating in a strategic environment that is geographically quite different, the navies of Europe often experience greater difficulty in arguing the case for a long-war capacity, and for the kind of naval forces appropriate to it, than does the US Navy. However, despite the fact that the inhabitants of West Germany or Denmark, and their political leaders, find a war possibly lasting for months a difficult eventuality to take seriously, most European navies do nevertheless tend to echo American views about the need for sustainability and about the requirement to defend NATO's sea lines of communication (SLOC) across the Atlantic. Both of these concepts are plainly based on an assumption that the war will last long enough for them to be a relevant consideration.

This is interesting because it suggests that the differences between a distinctively national American maritime strategy and equivalent NATO formulations are more apparent than real. Since the whole structure of NATO rests on the view that the strategic interests of the United States and Western Europe are at least complementary if not identical, it would be surprising, and indeed worrying, if this were not so.

From this perspective West Europeans will tend to support the ideas in the Maritime Strategy for two basic reasons. Firstly, anything which makes their large ally feel more secure, should also contribute to the credibility of the protection it provides for them. In such a case it is at least arguable that they would benefit indirectly, if, for example, the US Navy (with or without their help) were able to threaten a change in 'the nuclear correlation of forces' by an effective anti-SSBN campaign.[7] Secondly, they should often find that they too would benefit from a particular strategic option, though possibly not for quite the same reasons as the Americans. Thus, for example, they might profit most from an anti-SSBN campaign, not in some arcane strategic nuclear way, but because such a strategy would force the Soviet Union to protect its SSBNs with maritime assets that would otherwise be free to attack European maritime interests somewhere else.

European navies are therefore likely to find themselves broadly sympathetic to the aims of the Maritime Strategy and engaged in the same kind of operations, at least to the extent they can, as their larger American ally. It is perhaps worth repeating the point that if this were not so, NATO would be in very serious trouble.

The basic similarity of the strategic interests of the US and West European navies is also suggested by the apparent resemblances between the Maritime Strategy, and the equivalent NATO formulation, the Concept of Maritime Operations (CONMAROPS). While analysis is constrained by the fact that CONMAROPS is a classified document produced by the three major NATO Commanders (namely, SACEUR, SACLANT and CINCHAN), its broad contents are public knowledge. They were described thus by Admiral Harry Train, then SACLANT, in 1981:[8]

> This strategy involves sealing off the Soviet Navy in their home waters; the Northern Fleet as far north in the Norwegian Sea as possible, the Baltic Fleet in the Baltic and the Black Sea Fleet in the Black Sea — it is easier to do those latter two than the former. At the same time, we aim to move our shipping across the Atlantic as far south as possible so that they cross the Atlantic roughly at the tropic of Cancer...all the way over to an area south of ...Madeira...where they would turn north and go up to the south-west approaches to the English Channel. This keeps the ships out of range of Soviet land-based air for the maximum time possible and permits us to concentrate our escorts on that shorter portion of the transit...

The similarities between this kind of thinking and that displayed in the Maritime Strategy are obvious, and, given the fact that they are intended to counter the same hostile navy in the same geographic circumstances, not at all surprising.

A STRATEGY OF GLOBALISM

The global perspectives of the Maritime Strategy, together with the growing interest of the United States in the Pacific Ocean, could possibly pose European navies in the Northern Flank area with a number of issues to be resolved. The first, touched on in the previous section, is the prospect of European countries and their navies being involved in conflict off Europe's Northern Waters as a result of some distant conflict in the Third World spreading to Europe by a process of horizontal escalation.

The unattractiveness of this prospect persuades some of the wisdom of joining in American endeavours in the outside world, perhaps partly to moderate American policy or to help contain local conflicts. As Michael Howard has pointed out, ' The ability to provide or withhold direct military cooperation in a planned intervention is the most direct way in which we could influence American policy at critical moments.[9]

There are other reasons for such participation, however. Being actively supportive of American policy in this way might prove a small insurance premium if it helps bolt American power more securely to the defence of Western Europe. Many European countries, moreover, have their own reasons for a presence out-of-area.

Typically it is navies which put this point of view and Admiral Hulshof of the Netherlands Navy spoke for most of them when he remarked recently that conflicts are much more likely to occur outside Europe than inside it and the NATO boundary 'was never meant to mark the limits of our interests...' which would be at risk '...unless we are prepared to meet the challenge at sea and build and maintain a naval deterrent capable of operations outside the boundary'.[10] The British, French, Italians and Dutch have the capacity and have frequently exercised it in recent years, most notably in the mine-clearance exercise in the Red Sea in 1984, in operations off the Lebanon, and currently in the Gulf.

However, whether the ships in question belong to the US Navy or to those of the blue-water fleets of the West Europeans themselves, they cannot be in two places at once. Interestingly, the Maritime Strategy itself contains tables which show how long it would take for carriers to get from places like the Indian Ocean to the Norwegian Sea in an emergency.

There would seem to be two implications of this globalism for naval operations in the Norwegian Sea. Firstly, the requirement for forces, especially but not exclusively American ones, to go outside the NATO area means they will be less able to contribute to that multilateralism, which we shall see shortly, is often regarded as an important ingredient of deterrence in Northern Waters. These centrifugal tendencies would be particularly damaging if they were conducted without warning or consultation. The need to avoid this was re-emphasised by General Rogers after the Bonn Declaration of 1982:

This has been deemed especially important if any nation is considering an out-of-area deployment of forces, so that the effect of such a deployment on Alliance security and defence capabilities can be examined collectively by the appropriate NATO bodies.[11]

The second implication is even more obvious. If American and other major European naval capabilities are involved in commitments out-of-area, the requirement for the locals in times of tension or war to hold the fort until they arrive may well grow. Although in some quarters there may be worries about American carrier battle groups being in the Norwegian Sea, in others there are fears that they might not be, or at least not in time.

A STRATEGY OF DETERRENCE

The Maritime Strategy has won the adherence of previous opponents like James Schlesinger, because the US Navy has now carefully avoided confirming earlier impressions of unilateralism.[12] The emphasis now is all on the fact that an essential component of a successful strategy of deterrence will be on supporting allies and, indeed, on winning their support. Since some of those allies, like Norway and Turkey, are geographically much in front of the main centres of Western military power, this support clearly requires forward deployments and operations. Doing anything else would be 'unacceptable, morally, legally and strategically. Allied strategy *must* be prepared to fight in forward areas. That is where our allies are and where our adversary will be.'[13]

When dealing with countries like Norway and Sweden, the Soviet Union's evident approach is to seek to 'bilateralise' its security relationships with them. The consequent relationship would obviously be an unequal one because the Soviet Union is so much stronger than they are; the end result of this policy, if it were successful, therefore, would be a degree of deference from the smaller power. The smaller power's only response in such a situation is to encourage its allies to move to its support so that the security issues are 'multilateralised' and the distribution of power is more even.

For these reasons most West Europeans welcome the forward deployment of American forces. In the naval sphere this is indicated by Allied involvement in a series of NATO exercises off and on the Northern Flank which have grown in scope over recent years, matching the increase in capacity of the Soviet Navy. In this connection it is worth making the point that the steady extension of the Soviet defensive perimeter means that more and more areas stand in danger of being cut off from their main centre of military support. In the summer exercise carried out by the Soviet Fleet in 1985, for example, the submarine picket line was established south of Ireland. For the same reasons, there is a support for the idea of creating a kind of standing naval force for the Norwegian Sea like STANAVFORLANT or at least NAVOCFORMED (forces maintained in the Atlantic and Mediterranean respectively). However, suggestions, made by the Dutch, that such a force be formed were delayed for the time being, lest it damage STANAVFORLANT. Instead, the idea is to deploy Allied forces, including STANAVFORLANT itself, more routinely into the Norwegian Sea.[14]

This tendency to take the situation in Northern Waters more seriously has been evident in many European navies. The British Navy, for example, now pays the area much more attention than it used to and conducts exercises up there, individually and in concert with its allies, on a regular basis. In 1980, the German Navy likewise was released from the self-imposed restraints which had hitherto limited its activities to 61° North, and is building the appropriate ships.

It would be wrong, however, to suppose that this process of naval multilateralisation was entirely without controversy. The point is often made that the purpose of naval forces is to prevent war, not precipitate it. There is therefore a need to reassure the Soviet Union that resolute acts of collective deterrence do not imply hostile intent. For this reason European naval forces are likely to become involved in efforts to evolve confidence-building measures, or even acts of arms control with the Soviet Union in Northern Waters, and it is not inconceivable that their interests in such matters may sometimes diverge a little. Whereas, for example, the Norwegians may well be interested in the idea of the prior notification of naval exercises, traditional sea powers like the United States, fiercely attached to the concept of the freedom of the seas, might be more hesitant.

In the same way, the vulnerability of ships, submarines, aircraft and sensors to sudden attack by naval missiles has increased naval commanders' incentives to win what Admiral Gorshkov has called 'the battle for the first salvo'. There is therefore a tendency for them to press for more permissive rules of engagement, and the extent to which they do so sometimes reflects different national perspectives on the matter.[15] Traditionally, the Americans have been particularly keen on protecting large and valuable assets like aircraft carriers from the danger of undeclared attack by methods bordering on the anticipatory.

Navies without such assets can afford to be a shade more relaxed.

The danger over all such naval professional matters as this is that they seep through to the body politic and become divisive, threatening the Allied cohesion, the need for which is so emphasised in the Maritime Strategy. The presentation of naval policy therefore matters not only in that it should not feed Soviet paranoia, but also that it should not cause perturbations amongst allies. 'Rambo' rhetoric should be avoided for military as well as political reasons.

THE MARITIME STRATEGY AS A STRATEGY

The Maritime Strategy, like CONMAROPS, does not specify what will be done, when and with what. It is an intellectual framework encompassing a whole range of options that might or might not be tried in the event of conflict in Northern Waters, depending on the circumstances. It is therefore worth examining those options, and how the navies of Europe might be involved in them.

Firstly, as we have seen, the United States' growing commitments outside the NATO area could possibly lead to a longer period during which the Europeans would need to hold the fort until the big battalions in the shape of US carrier battle groups arrive. This requirement is widely recognised amongst the European navies:

> It must be assumed that only limited US Navy forces would be available in the Eastern Atlantic at the outbreak of hostilities. European Navies, and in particular the Royal Navy, must therefore be ready to play a leading role in initial operations.[16]

While a comparison of respective orders of battle between the Soviet Northern and Baltic Fleets and the navies of the North Europeans makes the containment of the former by the latter seem a very ambitious undertaking, the situation may not be so hopeless as might at first sight appear. The traditionally cautious Soviet Navy would no doubt be reluctant to hazard too high a proportion of its most important assets before their principal enemy, the US Navy, appears on the scene. Although all Soviet military thought, the naval variety included, stresses the need for surprise attack and the gaining of positions of advantage as early in the conflict as possible, apprehension about what might happen immediately afterwards, if they lost too many assets in the process, might well deter them from taking more risks than they strictly need to. With their penchant for the detailed analysis of relevant history, they cannot but be aware of one of the lessons of the Norwegian campaign of 1940: though successful, this caused the Germans serious naval losses that turned out to be gravely embarrassing later in the year.

Moreover, the modern technology of naval conflict is often said to be such that the navy concerned with sea denial has important advantages over the navy concerned with sea control. It is far from inconceivable that initial European efforts in defensive warfare around the Norwegian and Danish coastlines, and an active submarine campaign further to the east, could prove surprisingly disruptive of any plans the Soviet Union might have. A particular problem, however, might be if it were necessary to reinforce Norway. The wartime reinforcement of Norway, before the US Navy arrived, would obviously be the most demanding operational requirement the European navies might have to face.

The second element in which European navies would doubtless be involved in such a situation would obviously be that defence of coastal waters to which reference has already been made. This is, and would remain even after the US Navy arrived, the responsibility of the navies of Norway, Denmark and West Germany. They have considerable expertise in the specialised type of warfare in which coastal submarines, mines, fast patrol boats and shore defences of various sorts exploit to the maximum the topographic and oceanographic characteristics of the Scandinavian coastline. When, to this force, are added modern attack aircraft like the German Navy's formidable Tornado force in Schleswig-Holstein, the resultant defence is not something to be taken lightly.

Clearly, the worst danger in this regard is the prospect of sudden amphibious attack, particularly in the Baltic Approaches, where some Warsaw Pact amphibious bases are only a few hours away from their presumed landing sites. To a degree, however, this danger is somewhat reduced by the tactical rather than operational, and still less strategic, thrust of Soviet thinking about amphibious warfare[17] and their tendency to regard it as axiomatic that the ground forces should be able to link up with landed amphibious forces within a very short space of time. Both of these considerations tend to limit possibilities more than might appear at first glance. However that may be, coping with such a possibility is and will remain the primary task of local European navies in Northern Waters.

Thirdly, there is the question of European involvement in barrier operations, whose aim is to so patrol the gap between Greenland, Iceland and the United Kingdom as to contain the bulk of Soviet forces north of that line, and exacting significant attrition of forces which seek to pass through it. Although this kind of operation attracts much less public interest than it used to, it remains an important element in the struggle for mastery of Northern Waters.

European involvement in this would be of two sorts. Firstly, the importance of land-based air power in the imposition of such a barrier means that European aircraft, and European air bases, such as Keflavik, would be quite vital to the success of this element of the policy, a fact of which the Soviet Union is no doubt fully aware.

Secondly, European navies with a serious open-water capability, such as the French, British, Dutch and to an extent the German navies, could expect to contribute to the surface and submarine forces required. The greater size of the US Navy, its experience in barrier operations and the salience of the American SOSUS anti-submarine detection system would all tend to make this an American-led campaign. The same would seem to go for the fourth and most publicised element of the Maritime Strategy, namely Forward Operations. There is a good deal to be said for the view that Forward Operations are particularly ill-defined in most contemporary strategic debate, not least because they are no more than a list of possible campaign options.

Attention seems to focus, however, on two aspects of the case, that is to say on anti-SSBN operations and on Forward Operations by US Navy carrier battle groups. As we have already seen, the British and French navies might participate in the first of these for a whole variety of political and operational reasons.

The Maritime Strategy clearly seeks to distance itself from earlier, wilder perceptions of what Forward Operations by carrier battle group could mean by remarking that 'it does not imply some immediate "Charge of the Light Brigade" attack on the Kola Peninsula or any other specific target'.[18] For the most obvious operational reasons, the Soviet subsurface and air threat would need to be reduced to what naval professionals would regard as manageable proportions before any such operation was undertaken. Admittedly, detailed putative exchange analysis might lead some to argue that Forward Operations by carrier are one way of doing just that!

Generally, though, most analysts believe a degree of preparatory work would need to be done first, and it is by no means inconceivable that European navies might be involved in this. To an extent they could contribute to this preparatory phase by Forward Operations by submarine, by the maintenance of aircraft ASW patrols and by contributing to the degradation of Soviet land air power. The British, and possibly the French and Dutch, could also contribute to the direct

ASW protection of carrier battle groups by ASW task forces of various sorts. The extent to which such things are rehearsed suggests that this prospect is taken seriously.

The reward for success in such operations would be that the oversea reinforcement and resupply of Western Europe as a whole, and of northern Europe in particular, would be that much more easily accomplished. Indeed, this is often taken to be the chief justification for engaging in Forward Operations in the first place.[19] The involvement of the navies of Europe in the protection of SLOCs across the Atlantic is so well known as hardly to need description, but it is sometimes forgotten that there is a more localised variant of this too. This is the despatch of reinforcements to Norway and Denmark, activities whose protection would most certainly require heavy North European involvement.

In conclusion, it seems fair to say that although the Maritime Strategy is often presented as basically an American affair, and indeed is sometimes explained away as no more than a budgetary device, it has many implications for, and clearly depends on, the navies of northern Europe to a much greater extent than meets the eye.

NOTES

1. The views expressed in this paper should not necessarily be taken to reflect official opinion in any way.
2. *The Maritime Strategy*, (US Naval Institute, January 1986). (hereinafter MS).
3. Ibid., pp. 4–5.
4. Ibid., p. 4.
5. S.G. Gorshkov, *The sea power of the state*, (Pergamon, London, 1979).
6. See, for example, Christopher Donnelly, 'The development of the Soviet concept of echelonning' *NATO Review*, December 1984.
7. MS, p. 14.
8. In Christopher Coker, *US military power in the 1980s* (Macmillan, London, 1984).
9. Michael Howard, remarks to House of Commons Select Committee on Foreign Affairs, 1980 quoted in P. Foot, 'Improving capabilities for extra-European contingencies: the British contribution', *ASIDES*, no 18 (Centre for Defence Studies, Aberdeen, Spring 1981), p. 26.
10. Vice Admiral J.H.B. Hulshof, 'The Royal Netherlands Navy on escort and ASW tasks', *NATO's fifteen nations*, Special Edition, 2 (1982).
11. Quoted in *ACE Output*, vol. 2, no. 5, (September 1982), pp. 16–17.
12. Robert S. Wood, in G. Till (ed.), *Britain and the security of the Northern Flank*, (Macmillan, London, 1988).
13. MS, p. 7.

14. Sir James Eberle in Till (ed.), *Britain*.
15. See, for example, 'NATO changing rules to allow sea commanders to fire first', *Daily Telegraph*, (London) 22 March 1984.
16. *Statement on the Defence Estimates 1985*, Cmnd 9430 (HMSO, London), p. 53.
17. See Christopher Donnelly *et al.*, *Soviet amphibious operations: implications for the security of NATO's Northern Flank*, (Soviet Studies Centre, Sandhurst, 1985).
18. MS, p. 10.
19. MS, p. 11.

Index

abyssal plains 71
air bases *see under individual countries and* NATO
Alaska 1, 16, 72
ALCM (air-launched cruise missile) *see under* United States; USSR
Aleutian Islands 17, 96, 134
Alpha Ridge/Mendeleyev Ridge 71
Amundsen Basin 71
Antarctica 16, 50, 68
anti-submarine warfare *see* NATO operations; US Navy Maritime Strategy; USSR
Archer, C. 1, 164
Arctic Basins 71
Arctic Ocean
  baselines 69
  continental shelf delimitation 65
  continental shelf outer limits 71–2
  freedom of navigation 65
  historic bays 71
  ice-covered areas 64
  islands legal regime 55, 67–8
  marine pollution 64
  sector claims 66
  SSBN operating area 228
Arctic Mid-ocean Ridge 71
Arctic scientific research 11, 62–3
Arctic shipping 62
Arkhangel(sk) 20, 21, 136
Arktika ice-breaker 62
Arktikugol trust 27
armed forces *see* NATO forces; *and see under individual countries;*
arms control
  Arctic 'SSBN sanctuary' 130, 230
  CBMs 11, 132, 246
  Nordic NWFZ 7, 8, 10, 130
  polar arms control 229
Astrakhan 83

Baffin Bay 1
Baikal-Amur mainline (BAM) railway 62
Baku 83
Baltic Fleet (USSR) *see under* USSR, Naval Fleets
Baltic Sea 11, 79, 83, 166, 175, 180, 197, 199
Baltic Approaches 168, 170, 180, 187, 248
Barentsburg 27
Barents Sea
  continental shelf delineation 5, 32–6, 44–53, 86-7
  energy exploration, Norway 80, 82
  energy exploration, USSR 82–7
  exclusive economic zone (EEZ) 47
  Faroes fisheries in 78
  fisheries 76–9
  'Grey Zone' Agreement 47, 52, 68, 76–9
  'incidents' 40–1, 64, 69, 79
  marine pollution control 53, 58, 87
  missile tests 58, 63
  Norwegian-Soviet Fisheries Commission 47, 76
  offshore installations, military uses of 57
  resources 23, 33
  United Kingdom fisheries in 78
Barents Trough 65, 67
Bear Island 29, 51, 54, 65, 66, 79, 80
Belgian Navy 180
Bering Sea 69
Bering Strait 62, 135, 229
  incidents 70
  US-Soviet maritime demarcation 70
Black Sea Fleet (USSR) *see under* USSR, Naval Fleets
Boconor consortium 40, 63, 85

Bornholm Island 166

Canada 1, 3, 172
  Canadian Air Sea Transportable (CAST) Brigade 172, 182, 194
  defence policy 229
  NATO and 172
  reinforcement of Norway 172, 182, 194
Canada Basin 71
Cape Taran 83
Carter, (President) 4, 206
Chernavin, V. 17
Chukchi Plateau 72
Chukchi Sea 70, 71
Churchill, R. 5, 44
conservation 11
continental shelf
  conventions on 45, 48, 51, 66
  disputes *see under* Barents Sea; Svalbard
  Malta-Libya case 49
crisis stability and escalation 187, 192, 195, 219-20
Cuban Missile Crisis 18

Davis Strait 1, 228
Denmark
  Allied exercises in 166
  defence policy 164, 242
  fisheries in Svalbard 80
  foreign bases in 166
  Greenland and *see* Greenland
  NATO and 165
  Navy 180, 248
  Nordic NWFZ 169
  nuclear weapons in 166
  public opinion 174
  reinforcement by NATO 9, 170
Denmark Strait 228
drillships, ice-class 82

East Greenland Case 65
East Siberian continental shelf 71, 72
East Siberian Sea 62, 71
EC (European Community)
  Svalbard and 80
Ejde 168

Elf Petroleum 82
Epstein, J. 223
equitable criteria 49
equidistance principle 52
escalation 185-6, 200
ESSO 82
exclusive economic zone (EEZ) *see under* Barents Sea

Faroes
  Barents Sea fisheries 78
  defence installations 168
  local assembly 169
  nuclear-free zone 175
  security 168
  Svalbard fisheries 80
Federal Republic of Germany 3
  Barents Sea fisheries 78
  energy equipment sales to USSR 85
  forward naval operations 183, 248
  Navy 180, 246, 249
Finland
  defence policy 164
  neutrality 165, 180, 198
  northern airfields 121
  Sweden and air defence cooperation 133
  USSR 165; offshore energy programme and 42, 86
Fisheries
  Barents Sea 76-9
  Svalbard 79-80
Fram Straits 1
France
  Barents Sea fisheries 78
  energy equipment sales to USSR 85
  Navy 198, 241, 244, 249
Franz Joseph Land 29, 37, 48, 52, 109
Friedman, N. 184

German Democratic Republic
  Barents Sea fisheries 78
  Svalbard fisheries 80
GIUK (Greenland-Iceland-UK) gap 8, 21, 91, 96, 108, 109-10, 117, 136, 138, 139, 140, 146, 147, 194

INDEX

Gorbachev, M. 10, 11
Gorshkov, S. 17, 18, 19, 241, 246
Greenland
    continental shelf delimitation 80
    defence areas and installations 168–9
    Denmark and 168
    fisheries 80
    Home Rule Assembly 169
    military importance 186, 228
    NATO and 168
    nuclear-free zone 175
    public opinion 175
    USSR and 80, 169
Greenland Sea
    military operations 138
    submarine operations 117
'Grey Zone' agreement *see under* Barents Sea
Grotius, H. 24
Guinea-Guinea Bissau maritime boundary arbitration 49
Gulf of Maine case 49, 51

Haltenbanken 82
Hammerfest 82
Harstadt 82
Hauge, S. 171
Holst, J. 4, 181
Hope Island 51
'horizontal escalation' *see under* United States Maritime Strategy
'Hotline' (US-Soviet) 201
Howard, M. 244
Hugemark, B. 128
Hulshof, J. 244

ice-breakers 3, 62
Iceland
    coastguard 170
    fisheries 171
    NATO and 164, 170
    nuclear-free zone 175
    People's Alliance Party 170
    public opinion 174
    security 3, 170–1
    threat perceptions 174
    USA and 167, 170, 171, 174
    whaling 171

Iceland Defence Force 167, 170
ice shelves, floes and islands 63
ice stations 63
INF (intermediate-range nuclear force) negotiations 175
internal waters 71
ICES (International Council on the Exploration of the Seas) 77, 81
ICJ (International Court of Justice) 45, 49, 57, 65
Incidents at Sea Agreement (US-USSR) 181
Ireland 246
Italian Navy 244

Jan Mayen Island 1
Jenisch, U. 5, 62
joint development zones 53, 65

Kamchatka Peninsula 94, 96, 134, 206
Kara Sea 40, 63, 71, 84, 117
Karadag 83
Kaufmann, W. 216
Keflavik air base 167, 170, 171, 174, 249
Kelley, P. 187, 207
Khrushchev, N. 94
Kirkenes 82, 86
Koivisto, M. 10
Kola Inlet 136
Kolguyev Island 82
Komer, R. 217, 223

Lansing, R. 26
Lappland, defence of 133
Lange, H. 167
Laptev Sea 71
Law of the Sea Convention (LOSC) 45, 55, 63, 64–6, 67, 71–2
Lehman, J. 184, 187, 206, 207, 208, 209
Lofoten Islands 107
Lomonosov Ridge 71
Longyearbyen 27
Loran-C navigation system 168
LOSC *see* Law of the Sea Convention

Luttwak, E. 218, 224

Mahan, A. 213
Makarov Basin 71
Malene Østervold research vessel 69
Mearsheimer, J. 216, 218, 219
median line 45, 48, 76
Miller, S. 9, 205
Mjørkadal air force station 168
Molotov, V. 30
Motzfeldt, J. 169
Murmansk 10, 40, 63, 240, 241
Murmansk Basin 84
Murmansk High structure 82
Murmansk Speech 10–11
Mustin, H. 208, 217

Nansen Basin 71
Narssarssuaq 168
NATO commands
  ACE (Allied Command Europe) 190
  AFNORTH (Allied Forces Northern Europe) 172, 229
  'Arctic Command' 229
  CINCHAN (C-in-C Channel) 83, 243
  SACEUR (Supreme Allied Commander Europe) 136, 183, 185, 189, 190, 193, 197, 243
  SACLANT (Supreme Allied Commander Atlantic) 136, 140, 145, 146, 170, 173, 183, 186, 194, 229, 243
NATO Defence Planning Committee 194
NATO exercises
  'Northern Wedding' 184, 191
  'Ocean Safari' 184, 191, 208
NATO forces
  AMF (ACE Mobile Force) 172, 192, 193
  basing areas compared 136–8
  comparison with Soviet Navy 134–58
  MARCONFORLANT (Maritime Contingency Force Atlantic) 173

NAVOCFORMED (Standing Naval Force Mediterranean) 246
  operating areas compared 139–40
  role of European navies 247–50
  ship class obsolescence compared 140–4
  STANAVFORCHAN (Standing Naval Force Channel) 191, 193
  STANAVFORLANT (Standing Naval Force Atlantic) 173, 191, 193, 246
  Striking Fleet Atlantic 173, 194, 208, 217
  weapons/sensor quality 143
NATO North Atlantic Council 194
NATO operations
  anti-submarine warfare (ASW) 120, 186, 199–200, 241, 249
  anti-submarine warfare barriers 96, 138, 147, 248–9
  air defence barriers 121, 138
  GIUK gap 138–9, 147, 248
  Iceland-based 121, 167, 170
  maritime surveillance 180, 192, 194, 196, 249
  naval rules of engagement 195, 246
  out-of-area 244–5
  SLOC (sea lines of communication) defence 7, 92, 109–10, 134, 136, 139, 145, 180, 184, 185, 186, 188, 189, 196, 197, 200, 201, 210, 224 226, 242, 250
NATO reinforcement of Northern Flank
  Denmark 193
  Norway 172–3, 181–2, 184, 193–6, 198–9, 248, 250
NATO strategy
  CONMAROPS (Concept of Maritime Operations) 183–6, 188, 189–90, 197, 243, 247
  flexible response 185–6, 188, 200
  nuclear escalation option 186

255

NATO systems
  AWACS (Airborne Warning and Control System) 170, 192
  NAEW&C (NATO Early Warning and Control System) 173
  NORCISS (NATO Joint Command, Control and Information System) 173
naval force comparisons 134–59
Naval Strategy *see under* NATO strategy; USSR; United States
navies *see* NATO forces; *and under individual countries*
Netherlands
  Navy 180, 244, 246, 249
  reinforcement of Denmark 170; Norway 172
non-encroachment principle 50, 51
Nordic Defence Union 164
Nordic Nuclear Weapon-Free Zone (NNWFZ) 7, 8, 10, 130, 169–70, 175
  Denmark and 169–70
  Norway and 175
  USSR and 10
Norsk Hydro 82
North Cape 50, 54, 66, 137, 146, 180
North Pole 62, 63
Northern Fleet (USSR) *see under* USSR, Naval Fleets
Norton, D. 9, 179
Norway
  air defence 172, 196
  air fields 121, 128, 173, 198
  Air Force 138, 172–3
  Barents Sea fisheries 76–9
  Canadian forces 172, 182, 194
  Collocated Operating Bases agreements 173
  defence policy 4, 171–3, 228
  defence spending 172
  EEZ declaration 76
  Finnmark, Allied exercises in 166, 173
  foreign bases in Norwegian territory 9, 32, 166, 312
  NATO and 132, 165–6, 174
  naval exercises 182
  Navy 138, 172, 180, 182, 248
  non-alignment, neutrality 164–5
  NNWFZ 175
  pre-stocking, prepositioning of equipment 131, 172, 173, 182, 202
  offshore energy exploration *see under* Barents Sea; USSR
  public opinion on defence 174–5
  reassurance and deterrence 4–5, 8, 31, 131–2
  reinforcement of 8, 131, 172–3, 181–2, 193–6, 198–9, 209, 226, 248
  Svalbard 25–31, 53–8
  Svalbard fisheries 79–80
  Troms area 128, 211
  Trondheim 211
  US and 131, 173, 184, 210–11
  USSR and 27–43, 44–59 passim, 65–70, 76–87, 131–3, 166, 171, 173, 175–6
Norwegian Petroleum Consultants 85
Norwegian Petroleum Directorate 69
Norwegian Sea, military importance of 94, 107, 108, 117, 130, 131, 138, 140, 145–6, 172, 179, 186, 199, 208, 245–6
Novaya Zemlya 33, 37, 48, 82, 109
nuclear threshold 221–2, 227

Okhotsk Sea 16, 69, 215
Olesen, K. 166
Orion P-3 C maritime surveillance aircraft 173
Ørvik, N. 167
Østreng, W. 130

Pechora Sea 82, 84
Peter the Great 13, 16, 20
Petrobaltic 83
Petropavlovsk 17, 21, 134
Poland and Barents Sea fisheries 78
Pomor-Nordic Trade Company 85

Port Arthur 17
Posen, B. 220
Professor Shtokman research vessel 82
Pyramiden 27
Rauma Repola 82
Reagan Administration 206, 208, 224, 227, 230, 239, 241
Record, J. 216
Ries, T. 7, 9, 90, 167, 180, 241
Rogers, B. 185, 244
Russia and
  Alaska 16, 17
  Antarctica 16
  Baltic 16, 21
  Black Sea 16, 21
  Crimean War 17
  Great Northern Expedition 16, 21, 24
  Spitzbergen 25–6
Rybachii Peninsula 76

Saga Petroleum 85
SALT (Strategic Arms Limitation Treaty) 227
Schlesinger, J. 245
Schleswig-Holstein 248
Schluter, P. 4
Scrivener, D. 6, 76
SDI (Strategic Defense Initiative) 92, 94, 105, 109, 130, 175, 239
Sea of Japan 16, 120
sector claim 29, 34, 45, 50, 66, 68, 70, 76, 87
Seaforth Maritime 82
Severodvinsk 136
Severomorsk 63
shelf semi-sub rig 82
Shetland Isles 1
Sjaastad, A. 171
SLCM (sea-launched cruise missile) *see under* United States; USSR
SLOC (sea-lines of communication) *see under* NATO; USSR
Sollie, F. 6, 13
Søndre Strømfjord 168
Sornfelli 168
SOSUS (sound ocean surveillance system) 170, 249
Spanish fisheries
  Barents Sea 78
  Svalbard 79, 80
SSBNs (strategic nuclear submarines) *see under* United States; USSR
SSNs (nuclear powered attack submarines) *see under* United States; USSR
Statoil 82
Store Norske Spitsbergen Kulkompani 27
submarines ridges 71–2
surveillance satellites 168
Svalbard Archipelago
  continental shelf issue 37–8, 53–8, 67–9
  disputes and treaty 5, 15, 25–31, 65, 66–7, 70
  economic zone issue 53–8
  fisheries 37, 55, 79–80
  FPZ (fisheries protection zone) 37, 57, 67, 79–80
  marine pollution and conservation 54, 57
  non-militarisation 27, 67
  offshore installations, military use of 54, 57
  offshore energy exploration 54, 57
  sovereignty 26–31
  territorial waters 55–57, 67
  Denmark and 67, 80
  European Community and 80
  Faroes and 80
  FRG and 67
  France and 67
  GDR and 80
  Spain and 79–80
  UK and 54, 67, 80
  US and 54, 67
  USSR-Norwegian relations concerning 6, 27–31, 36–8, 53–8, 66–8
Svalbard Passage 64
Svalbard Plateau 84
Sweden
  air defence and cooperation with Finland 133

airfields 121, 128
defence policy 165, 227
non-alignment and neutrality 198
USSR and 42, 86
Thibault, G. 216
Thorshaven 168
Thule 168
Timano-Pechora Basin 84
Treholt, A. 48
Tromsøflaket 80
Tsushima (Battle of) 17
Turner, S. 216

United Kingdom
  Air Force 173
  Barents Sea fisheries 78
  Iceland fisheries 171
  Navy 180, 183, 241, 244, 246, 247, 249
  reinforcement of Denmark 170
  Svalbard fisheries 80
  UK Mobile Force 170
  UK-Netherlands Amphibious Force 172, 182, 194
  USSR offshore programme and 85
United Nations Law of the Sea Convention *see* Law of the Sea Convention
United States
  air power 99–100, 107, 130, 168, 171, 220, 227
  BMEWS (ballistic missile early warning system) 168
  Chief of Naval Operations 187, 189, 190, 195, 207
  Coastguard 144
  DEW (distant early warning) Line 168
  Fleets: Atlantic (2nd) 136, 139
  Mediterranean (6th) 136
  Pacific (7th) 190, 229
  Greenland and 168–9
  Iceland and 121, 166–7, 170–1
  Marine Amphibious Brigade 131, 172, 182, 189, 194, 195, 211
  Marine Corps Commandant 187, 207

Maritime Strategy ('Forward Maritime Strategy') *see* US Maritime Strategy
mine warfare forces 190
National Command Authorities 190
Norway and 131, 173, 184, 196, 210, 211, 226
Norwegian Sea, peacetime naval deployments in 184, 226
SLBMs: Polaris 108
  Poseidon 91
  Trident 91, 105, 108, 220, 227
SLCMs (Tomahawk) 91, 105, 130
Space Command 168
SSN attack submarines 209, 211, 214, 219, 228
US Maritime Strategy 3, 4, 9–10, 140, 173, 184–202, 205–30, 239–50
  aircraft carriers 9, 94, 211, 216
  anti-air operations 212
  amphibious operations against Soviet territory 210, 212, 217
  ASW 96, 199–201, 211–2, 217
  Arctic under-ice ASW operations by SSNs 96, 209, 211, 224, 228, 230, 240
  attrition of Soviet SSBNs 9, 199–201, 211, 212, 214, 216–21, 224, 241–3, 249
  carrier air-strikes on Soviet territory 214, 216, 217, 224, 240, 249
  carrier battle groups 136, 139, 140, 145, 190, 196, 212, 216–7, 221–2, 241, 245, 247, 249
  CONMAROPS and *see* NATO strategy
  continuity and innovation in US naval policy 213–5
  crisis management and 219
  criticisms of 215–25
  domestic (US) debate on 205–9, 214–5

European navies' role in 239–50
European perceptions of 207, 214
forward operations 209–17, 221–2, 225, 245, 249–50
globalism and 244–5
horizontal escalation 196–7, 240, 244
inadvertent escalation potential 219–22
intra-crisis deterrence 187–9, 191–7, 211, 219
northern theatre of maritime operations and 191–202, 225–30
nuclear 'correlation of forces' and 210, 211, 218, 220, 242
operations in Barents Sea 211, 215, 217, 219–20, 224
political control and 198–9, 219
sea control 211–2
war-fighting 189–91, 197–8, 210
war-termination 218, 220
USSR
airborne assault operations 128
aircraft carrier programme 110
amphibious assault operations 128, 138, 248
anti-carrier warfare 107, 222
anti-submarine warfare:
anti-SSBN 91, 105–7, 108, 117
pro-SSBN 140, 200, 201
tactical 146
aviation: anti-ship missiles 143, 212
Frontal Aviation 128
Long-Range Strategic Aviation 91, 108–9
ALCMs 109, 169
Arctic forward airfields 109, 117
bombers 109
Naval Aviation *see* USSR, Naval Aviation
coastal defence 137, 144, 146
Finland and *see under* Finland

fisheries: Baltic Sea 79
Barents Sea 76–9
Barents Sea FPZ 69
Greenland 80
Svalbard 79–80
Greenland and 80, 169
Ground Forces: North-Western TVD, operations in 120–9, 132
ICBM vulnerability 218, 220, 227
KGB maritime border guards 145
Leningrad Military District 100, 116, 121, 131, 180
Marines 137
maritime jurisdiction: baselines 69
continental shelf decree 66, 70
EEZ decree 66, 70
historic bays 71
internal waters 71
sector line 29, 34, 45, 50, 66, 68, 70, 76, 87
territorial sea 69
military strategy in Northern Waters 90–133: strategic support missions 116–120, 129, 130
theatre, front-level objectives 120–8
mine warfare 137, 140, 146
Naval Aviation 110, 117, 121, 146, 190, 212, 215, 243:
Backfire 'B' 110, 212
carrier-based 121
naval construction 141–2, 145
Naval Fleets: Baltic 117, 136, 192, 194
Black Sea 136, 243
Northern *see* USSR, Northern Fleet
Pacific 14, 184
naval strategy: forward line of defence 107, 108, 117, 121
response to US Navy Maritime Strategy 10, 222, 225
sea control mission 92, 108, 110, 116, 117, 130

sea denial mission 107, 108, 110, 116, 117, 121, 130
tactical nuclear weapons release 10, 220–2
see also under USSR, anti-submarine warfare, SLOC interdiction, SSBN
naval surface forces *see under* USSR, Northern Fleet, SLOC interdiction
Nordic Nuclear-Free Zone and 10, 175
Northern Fleet 14, 22, 64, 91, 105, 107, 108, 110, 117, 120, 134–59, 171, 184, 192, 194, 243:
bases 136
  exercises 140, 194, 226, 246
  force comparisons 134–59
  operating areas 139–40
  ship class obsolescence 140–5
  warship size 144–5
  weapons/sensor quality 143
Norway and: in Barents Sea 5–6, 32–6, 44–53, 65–6, 68–9, 84–7
  and *see under* Barents Sea
in Svalbard 5–6, 28–31, 37–9, 53–8, 66–8 and *see under* Svalbard Archipelago
oceanic theatres of military operations (OTVDs) 115, 116, 117, 120, 121, 129
oceanographic research 13, 20
offshore energy capabilities and technology base 40, 82–3
offshore energy cooperation in Barents Sea 11, 39–43, 63, 69, 84–7
  Canada and 85
  Finland and 42, 82, 86
  FRG and 85
  Norway and 82–7
  Sweden and 42, 85
  UK and 85
offshore energy exploration:
  Baltic Sea 83
  Barents and Kara Seas 81–2, 84–7

Black Sea 83
Caspian Sea 83
Okhotsk Sea 83
SLOC interdiction mission 10, 92, 109–10, 117, 129, 145, 146, 179, 184, 190, 196, 200, 201, 211
SSBNs: defence of, *see under* USSR, anti-submarine warfare
– pro-SSBN employment strategy 92–9, 227–8
operating areas 9, 91, 94, 105, 108, 117, 211, 221, 226, 227–8
Arctic Ocean operating area 228
submarines, attack 110, 140, 146, 190, 200, 211
submarines, strategic (SSB, SSBN types) 92–6, 99, 117, 129
submarine-launched ballistic missiles (SLBM) 92–9, 130, 227
Svalbard and 5–6, 28–31, 37–8, 53–8, 66–8 and *see under* Svalbard Archipelago
Sweden and *see under* Sweden
Theatre of Military Operations (TVD), North-Western 105, 115, 116, 128, 129, 133
United Kingdom and *see under* United Kingdom
USSR air defence:
airborne early warning aircraft 99, 121
Air Defence Forces (Voiska PVO) 116, 121, 130
Arkhangelsk Air Defence Sector 91, 100, 121
BMEW radars 105, 117, 134
forward airfields, seizure of 121–2, 128, 129
interceptor aircraft, types of 99, 121
regional air defence 116
strategic air defence 91, 99–100, 105

Valentin Shashin drillship 82
van Tol, R. 7, 8, 9, 134, 167
Varangerfjord 48
Vyborg 82

Watkins, J. 187, 195, 207, 208, 209, 211, 212, 217
West, F. 215
West Siberian Basin 84
Western Europe:

US Navy Maritime Strategy and 10, 239–50
USSR and 10–11
White Sea 136
Willoch, K. 182
World War Two 164, 174, 199, 242, 248

Yamal Peninsula 105
Yamburg gas field 84, 86

For Product Safety Concerns and Information please contact our EU
representative GPSR@taylorandfrancis.com
Taylor & Francis Verlag GmbH, Kaufingerstraße 24, 80331 München, Germany

www.ingramcontent.com/pod-product-compliance
Lightning Source LLC
Chambersburg PA
CBHW071812300426
44116CB00009B/1284